MW00606969

# INTERNATIONAL
# PROPERTY
# MAINTENANCE
## CODE®

*A Member of the International Code Family®*

# IPMC

Become a **Building Safety Professional Member** and Learn More about the Code Council

**GO TO WWW.ICCSAFE.ORG** for All Your Technical and Professional Needs Including:

> Codes, Standards and Guidelines

> Membership Benefits

> Education and Certification

> Communications on Industry News

2012 International Property Maintenance Code®

First Printing: April 2011
Second Printing: February 2012

ISBN: 978-1-60983-056-4 (soft-cover edition)

PRINTED IN THE U.S.A.

# PREFACE

## Introduction

Internationally, code officials recognize the need for a modern, up-to-date property maintenance code governing the maintenance of existing buildings. The *International Property Maintenance Code*®, in this 2012 edition, is designed to meet this need through model code regulations that contain clear and specific property maintenance requirements with required property improvement provisions.

This 2012 edition is fully compatible with all of the *International Codes*® (I-Codes®) published by the International Code Council (ICC)®, including the *International Building Code*®, *International Energy Conservation Code*®, *International Existing Building Code*®, *International Fire Code*®, *International Fuel Gas Code*®, *International Green Construction Code*™ (to be available March 2012), *International Mechanical Code*®, ICC *Performance Code*®, *International Plumbing Code*®, *International Private Sewage Disposal Code*®, *International Residential Code*®, *International Swimming Pool and Spa Code*™ (to be available March 2012), *International Wildland-Urban Interface Code*® and *International Zoning Code*®.

The *International Property Maintenance Code* provisions provide many benefits, among which is the model code development process that offers an international forum for code officials and other interested parties to discuss performance and prescriptive code requirements. This forum provides an excellent arena to debate proposed revisions. This model code also encourages international consistency in the application of provisions.

## Development

The first edition of the *International Property Maintenance Code* (1998) was the culmination of an effort initiated in 1996 by a code development committee appointed by ICC and consisting of representatives of the three statutory members of the International Code Council at that time, including: Building Officials and Code Administrators International, Inc. (BOCA), International Conference of Building Officials (ICBO) and Southern Building Code Congress International (SBCCI). The committee drafted a comprehensive set of regulations for existing buildings that was consistent with the existing model property maintenance codes at the time. This 2012 edition presents the code as originally issued, with changes reflected through the previous 2006 editions and further changes developed through the ICC Code Development Process through 2010. A new edition of the code is promulgated every three years.

This code is founded on principles intended to establish provisions consistent with the scope of a property maintenance code that adequately protects public health, safety and welfare; provisions that do not unnecessarily increase construction costs; provisions that do not restrict the use of new materials, products or methods of construction; and provisions that do not give preferential treatment to particular types or classes of materials, products or methods of construction.

## Adoption

The *International Property Maintenance Code* is available for adoption and use by jurisdictions internationally. Its use within a governmental jurisdiction is intended to be accomplished through adoption by reference in accordance with proceedings established in the jurisdiction's laws. At the time of adoption, jurisdictions should insert the appropriate information in provisions requiring specific local information, such as the name of the adopting jurisdiction. These locations are shown in bracketed words in small capital letters in the code and in the sample ordinance. The sample adoption ordinance on page xiii addresses several key elements of a code adoption ordinance, including the information required for insertion into the code text.

# Maintenance

The *International Property Maintenance Code* is kept up to date through the review of proposed changes submitted by code enforcing officials, industry representatives, design professionals and other interested parties. Proposed changes are carefully considered through an open code development process in which all interested and affected parties may participate.

The contents of this work are subject to change both through the Code Development Cycles and the governmental body that enacts the code into law. For more information regarding the code development process, contact the Codes and Standards Development Department of the International Code Council.

While the development procedure of the *International Property Maintenance Code* ensures the highest degree of care, ICC, its membership and those participating in the development of this code do not accept any liability resulting from compliance or noncompliance with the provisions because ICC does not have the power or authority to police or enforce compliance with the contents of this code. Only the governmental body that enacts the code into law has such authority.

# Code Development Committee Responsibilities (Letter Designations in Front of Section Numbers)

In each code development cycle, proposed changes to this code are considered at the Code Development Hearings by the International Property Maintenance/Zoning Code Development Committee, whose action constitutes a recommendation to the voting membership for final action on the proposed changes. Proposed changes to a code section having a number beginning with a letter in brackets are considered by a different code development committee. For example, proposed changes to code sections that have the letter [F] in front of them (e.g., [F] 704.1) are considered by the International Fire Code Development Committee at the Code Development Hearings.

The content of sections in this code that begin with a letter designation is maintained by another code development committee in accordance with the following:

[A]  =  Administrative Code Development Committee;

[F]  =  International Fire Code Development Committee;

[P]  =  International Plumbing Code Development Committee; and

[B]  =  International Building Code Development Committee (IBC—Fire Safety, General, Means of Egress or Structural);

Note that, for the development of the 2015 edition of the I-Codes, there will be two groups of code development committees and they will meet in separate years. The groupings are as follows:

| Group A Codes (Heard in 2012, Code Change Proposals Deadline: January 3, 2012) | Group B Codes (Heard in 2013, Code Change Proposals Deadline: January 3, 2013) |
|---|---|
| International Building Code | Administrative Provisions (Chapter 1 all codes except IRC and ICC PC, administrative updates to currently referenced standards, and designated definitions) |
| International Fuel Gas Code | International Energy Conservation Code |
| International Mechanical Code | International Existing Building Code |
| International Plumbing Code | International Fire Code |
| International Private Sewage Disposal Code | International Green Construction Code |
| | ICC Performance Code |
| | International Property Maintenance Code |
| | International Residential Code |
| | International Swimming Pool and Spa Code |
| | International Wildland-Urban Interface Code |
| | International Zoning Code |

Code change proposals submitted for code sections that have a letter designation in front of them will be heard by the respective committee responsible for such code sections. Because different committees will meet in different years, it is possible that some proposals for this code will be heard by a committee in a different year than the year in which the primary committee for this code meets.

For instance, Section 502.1 is designated as the responsibility of the International Plumbing Code Development Committee, along with most of the provisions in Chapter 5. This committee will meet in 2012 to consider all code change proposals to the *International Plumbing Code* and any portions of other codes that it is responsible for, including Section 502.1 and most of the provisions of Chapter 5 (designated with [P] in front of those sections.) Therefore, any proposals to Section 502.1 in Chapter 5 will be needed to be submitted by January 3, 2012, for consideration in 2012 by the International Plumbing Code Committee.

Note that every section of Chapter 1 of this code is designated as the responsibility of the Administrative Code Development Committee, and that committee is part of the Group B portion of the hearings. This committee will hold its code development hearing in 2013 to consider all code change proposals for Chapter 1 of this code and proposals for Chapter 1 of all I-Codes except the *International Residential Code* and ICC *Performance Code*. Therefore, any proposals received for Chapter 1 of this code will be assigned to the Administrative Code Development Committee for consideration in 2013.

It is very important that anyone submitting code change proposals understand which code development committee is responsible for the section of the code that is the subject of the code change proposal. For further information on the code development committee responsibilities, please visit the ICC web site at www.iccsafe.org/scoping.

# Marginal Markings

Solid vertical lines in the margins within the body of the code indicate a technical change from the requirements of the previous edition. Deletion indicators in the form of an arrow ( ➡ ) are provided in the margin where an entire section, paragraph, exception or table has been deleted or an item in a list of items or a table has been deleted.

# Italicized Terms

Selected terms set forth in Chapter 2, Definitions, are italicized where they appear in code text. Such terms are not italicized where the definition set forth in Chapter 2 does not impart the intended meaning in the use of the term. The terms selected have definitions which the user should read carefully to facilitate better understanding of the code.

# EFFECTIVE USE OF THE INTERNATIONAL PROPERTY MAINTENANCE CODE

The *International Property Maintenance Code* (IPMC) is a model code that regulates the minimum maintenance requirements for existing buildings.

The IPMC is a maintenance document intended to establish minimum maintenance standards for basic equipment, light, ventilation, heating, sanitation and fire safety. Responsibility is fixed among owners, operators and occupants for code compliance. The IPMC provides for the regulation and safe use of existing structures in the interest of the social and economic welfare of the community.

## Arrangement and Format of the 2009 IPMC

Before applying the requirements of the IPMC it is beneficial to understand its arrangement and format. The IPMC, like other codes published by ICC, is arranged and organized to follow sequential steps that generally occur during an inspection. The IPMC is divided into eight different parts:

| Chapters | Subjects |
|----------|----------|
| 1 | Administration |
| 2 | Definitions |
| 3 | General Requirements |
| 4 | Light, Ventilation and Occupancy Limitations |
| 5 | Plumbing Facilities and Fixture Requirements |
| 6 | Mechanical and Electrical Requirements |
| 7 | Fire Safety Requirements |
| 8 | Referenced Standards |

The following is a chapter-by-chapter synopsis of the scope and intent of the provisions of the *International Property Maintenance Code:*

**Chapter 1 Scope and Administration.** This chapter contains provisions for the application, enforcement and administration of subsequent requirements of the code. In addition to establishing the scope of the code, Chapter 1 identifies which buildings and structures come under its purview. Chapter 1 is largely concerned with maintaining "due process of law" in enforcing the property maintenance criteria contained in the body of the code. Only through careful observation of the administrative provisions can the building official reasonably expect to demonstrate that "equal protection under the law" has been provided.

**Chapter 2 Definitions.** All terms that are defined in the code are listed alphabetically in Chapter 2. While a defined term may be used in one chapter or another, the meaning provided in Chapter 2 is applicable throughout the code.

Where understanding of a term's definition is especially key to or necessary for understanding of a particular code provision, the term is shown in italics wherever it appears in the code. This is true only for those terms that have a meaning that is unique to the code. In other words, the generally understood meaning of a term or phrase might not be sufficient or consistent with the meaning prescribed by the code; therefore, it is essential that the code-defined meaning be known.

Guidance regarding tense, gender and plurality of defined terms as well as guidance regarding terms not defined in this code is provided.

**Chapter 3 General Requirements.** Chapter 3, "General Requirements," is broad in scope. It includes a variety of requirements for the exterior property areas as well as the interior and exterior elements of the structure. This chapter provides requirements that are intended to maintain a minimum level of safety and sanitation for both the general public and the occupants of a structure, and to maintain a building's structural and weather-resistance performance. Chapter 3 provides specific criteria for regulating the installation and maintenance of specific building components; maintenance requirements for vacant structures and land; requirements regulating the safety, sanitation and appearance of the interior and exterior of structures and all exterior property areas; accessory structures; vehicle storage regulations and establishes who is responsible for complying with the chapter's provisions. This chapter also contains the requirements for swimming pools, spas and hot tubs and the requirements for protective barriers and gates in these barriers. Chapter 3 establishes the responsible parties for exterminating insects and rodents, and maintaining sanitary conditions in all types of occupancies.

**Chapter 4 Light, Ventilation and Occupancy Limitations.** The purpose of Chapter 4 is to set forth these requirements in the code and to establish the minimum environment for occupiable and habitable buildings, by establishing the minimum criteria for light and ventilation and identifies occupancy limitations including minimum room width and area, minimum ceiling height and restrictions to prevent overcrowding. This chapter also provides for alternative arrangements of windows and other devices to comply with the requirements for light and ventilation and prohibits certain room arrangements and occupancy uses.

**Chapter 5 Plumbing Facilities and Fixture Requirements.** Chapter 5 establishes the minimum criteria for the installation, maintenance and location of plumbing systems and facilities, including the water supply system, water heating appliances, sewage disposal system and related plumbing fixtures.

Sanitary and clean conditions in occupied buildings are dependent upon certain basic plumbing principles, including providing potable water to a building, providing the basic fixtures to effectively utilize that water and properly removing waste from the building. Chapter 5 establishes the minimum criteria to verify that these principles are maintained throughout the life of a building.

**Chapter 6 Mechanical and Electrical Requirements.** The purpose of Chapter 6 is to establish minimum performance requirements for heating, electrical and mechanical facilities and to establish minimum standards for the safety of these facilities.

This chapter establishes minimum criteria for the installation and maintenance of the following: heating and air-conditioning equipment, appliances and their supporting systems; water-heating equipment, appliances and systems; cooking equipment and appliances; ventilation and exhaust equipment; gas and liquid fuel distribution piping and components; fireplaces and solid fuel-burning appliances; chimneys and vents; electrical services; lighting fixtures; electrical receptacle outlets; electrical distribution system equipment, devices and wiring; and elevators, escalators and dumbwaiters.

**Chapter 7 Fire Safety Requirements.** The purpose of Chapter 7 is to address those fire hazards that arise as the result of a building's occupancy. It also provides minimum requirements for fire safety issues that are most likely to arise in older buildings.

This chapter contains requirements for means of egress in existing buildings, including path of travel, required egress width, means of egress doors and emergency escape openings.

Chapter 7 establishes the minimum requirements for fire safety facilities and fire protection systems, as these are essential fire safety systems.

**Chapter 8 Referenced Standards.** The code contains numerous references to standards that are used to regulate materials and methods of construction. Chapter 8 contains a comprehensive list of all standards that are referenced in the code. The standards are part of the code to the extent of the reference to the standard. Compliance with the referenced standard is necessary for compliance with this code. By providing specifically adopted standards, the construction and installation requirements necessary for compliance with the code can be readily determined. The basis for code compliance is, therefore, established and available on an equal basis to the code official, contractor, designer and owner.

Chapter 8 is organized in a manner that makes it easy to locate specific standards. It lists all of the referenced standards, alphabetically, by acronym of the promulgating agency of the standard. Each agency's standards are then listed in either alphabetical or numeric order based upon the standard identification. The list also contains the title of the standard; the edition (date) of the standard referenced; any addenda included as part of the ICC adoption; and the section or sections of this code that reference the standard.

# LEGISLATION

The *International Codes* are designed and promulgated to be adopted by reference by legislative action. Jurisdictions wishing to adopt the 2012 *International Property Maintenance Code* as an enforceable regulation governing existing structures and premises should ensure that certain factual information is included in the adopting legislation at the time adoption is being considered by the appropriate governmental body. The following sample adoption legislation addresses several key elements, including the information required for insertion into the code text.

## SAMPLE LEGISLATION FOR ADOPTION OF THE
## *INTERNATIONAL PROPERTY MAINTENANCE CODE*
## ORDINANCE NO._____

A[N] [ORDINANCE/STATUTE/REGULATION] of the [JURISDICTION] adopting the 2012 edition of the *International Property Maintenance Code*, regulating and governing the conditions and maintenance of all property, buildings and structures; by providing the standards for supplied utilities and facilities and other physical things and conditions essential to ensure that structures are safe, sanitary and fit for occupation and use; and the condemnation of buildings and structures unfit for human occupancy and use, and the demolition of such existing structures in the [JURISDICTION]; providing for the issuance of permits and collection of fees therefor; repealing [ORDINANCE/STATUTE/REGULATION] No. _____ of the [JURISDICTION] and all other ordinances or parts of laws in conflict therewith.

The [GOVERNING BODY] of the [JURISDICTION] does ordain as follows:

**Section 1.** That a certain document, three (3) copies of which are on file in the office of the [TITLE OF JURISDICTION'S KEEPER OF RECORDS] of [NAME OF JURISDICTION], being marked and designated as the *International Property Maintenance Code*, 2012 edition, as published by the International Code Council, be and is hereby adopted as the Property Maintenance Code of the [JURISDICTION], in the State of [STATE NAME] for regulating and governing the conditions and maintenance of all property, buildings and structures; by providing the standards for supplied utilities and facilities and other physical things and conditions essential to ensure that structures are safe, sanitary and fit for occupation and use; and the condemnation of buildings and structures unfit for human occupancy and use, and the demolition of such existing structures as herein provided; providing for the issuance of permits and collection of fees therefor; and each and all of the regulations, provisions, penalties, conditions and terms of said Property Maintenance Code on file in the office of the [JURISDICTION] are hereby referred to, adopted, and made a part hereof, as if fully set out in this legislation, with the additions, insertions, deletions and changes, if any, prescribed in Section 2 of this ordinance.

**Section 2.** The following sections are hereby revised:

Section 101.1. Insert: [NAME OF JURISDICTION]

Section 103.5. Insert: [APPROPRIATE SCHEDULE]

Section 112.4. Insert: [DOLLAR AMOUNT IN TWO LOCATIONS]

Section 302.4. Insert: [HEIGHT IN INCHES]

Section 304.14. Insert: [DATES IN TWO LOCATIONS]

Section 602.3. Insert: [DATES IN TWO LOCATIONS]

Section 602.4. Insert: [DATES IN TWO LOCATIONS]

**Section 3.** That [ORDINANCE/STATUTE/REGULATION] No. _____ of [JURISDICTION] entitled [FILL IN HERE THE COMPLETE TITLE OF THE LEGISLATION OR LAWS IN EFFECT AT THE PRESENT TIME SO THAT THEY WILL BE REPEALED BY DEFINITE MENTION] and all other ordinances or parts of laws in conflict herewith are hereby repealed.

**Section 4.** That if any section, subsection, sentence, clause or phrase of this legislation is, for any reason, held to be unconstitutional, such decision shall not affect the validity of the remaining portions of this ordinance. The [GOVERNING BODY] hereby declares that it would have passed this law, and each section, subsection, clause or phrase thereof, irrespective of the fact that any one or more sections, subsections, sentences, clauses and phrases be declared unconstitutional.

**Section 5.** That nothing in this legislation or in the Property Maintenance Code hereby adopted shall be construed to affect any suit or proceeding impending in any court, or any rights acquired, or liability incurred, or any cause or causes of action acquired or existing, under any act or ordinance hereby repealed as cited in Section 3 of this law; nor shall any just or legal right or remedy of any character be lost, impaired or affected by this legislation.

**Section 6.** That the **[JURISDICTION'S KEEPER OF RECORDS]** is hereby ordered and directed to cause this legislation to be published. (An additional provision may be required to direct the number of times the legislation is to be published and to specify that it is to be in a newspaper in general circulation. Posting may also be required.)

**Section 7.** That this law and the rules, regulations, provisions, requirements, orders and matters established and adopted hereby shall take effect and be in full force and effect **[TIME PERIOD]** from and after the date of its final passage and adoption.

# TABLE OF CONTENTS

# CHAPTER 1

# SCOPE AND ADMINISTRATION

## PART 1 — SCOPE AND APPLICATION

### SECTION 101
### GENERAL

**[A] 101.1 Title.** These regulations shall be known as the *International Property Maintenance Code* of [NAME OF JURISDICTION], hereinafter referred to as "this code."

**[A] 101.2 Scope.** The provisions of this code shall apply to all existing residential and nonresidential structures and all existing *premises* and constitute minimum requirements and standards for *premises*, structures, equipment and facilities for light, *ventilation*, space, heating, sanitation, protection from the elements, life safety, safety from fire and other hazards, and for safe and sanitary maintenance; the responsibility of *owners*, *operators* and *occupants*; the *occupancy* of existing structures and *premises*, and for administration, enforcement and penalties.

**[A] 101.3 Intent.** This code shall be construed to secure its expressed intent, which is to ensure public health, safety and welfare insofar as they are affected by the continued *occupancy* and maintenance of structures and *premises*. Existing structures and *premises* that do not comply with these provisions shall be altered or repaired to provide a minimum level of health and safety as required herein.

**[A] 101.4 Severability.** If a section, subsection, sentence, clause or phrase of this code is, for any reason, held to be unconstitutional, such decision shall not affect the validity of the remaining portions of this code.

### SECTION 102
### APPLICABILITY

**[A] 102.1 General.** Where there is a conflict between a general requirement and a specific requirement, the specific requirement shall govern. Where differences occur between provisions of this code and the referenced standards, the provisions of this code shall apply. Where, in a specific case, different sections of this code specify different requirements, the most restrictive shall govern.

**[A] 102.2 Maintenance.** Equipment, systems, devices and safeguards required by this code or a previous regulation or code under which the structure or *premises* was constructed, altered or repaired shall be maintained in good working order. No *owner*, *operator* or *occupant* shall cause any service, facility, equipment or utility which is required under this section to be removed from or shut off from or discontinued for any occupied dwelling, except for such temporary interruption as necessary while repairs or alterations are in progress. The requirements of this code are not intended to provide the basis for removal or abrogation of fire protection and safety systems and devices in existing structures. Except as otherwise specified herein, the *owner* or the *owner's* designated agent shall be responsible for the maintenance of buildings, structures and *premises*.

**[A] 102.3 Application of other codes.** Repairs, additions or alterations to a structure, or changes of *occupancy*, shall be done in accordance with the procedures and provisions of the *International Building Code*, *International Energy Conservation Code*, *International Fire Code*, *International Fuel Gas Code*, *International Mechanical Code*, *International Residential Code*, *International Plumbing Code* and NFPA 70. Nothing in this code shall be construed to cancel, modify or set aside any provision of the *International Zoning Code*.

**[A] 102.4 Existing remedies.** The provisions in this code shall not be construed to abolish or impair existing remedies of the jurisdiction or its officers or agencies relating to the removal or demolition of any structure which is dangerous, unsafe and insanitary.

**[A] 102.5 Workmanship.** Repairs, maintenance work, alterations or installations which are caused directly or indirectly by the enforcement of this code shall be executed and installed in a *workmanlike* manner and installed in accordance with the manufacturer's instructions.

**[A] 102.6 Historic buildings.** The provisions of this code shall not be mandatory for existing buildings or structures designated as historic buildings when such buildings or structures are judged by the *code official* to be safe and in the public interest of health, safety and welfare.

**[A] 102.7 Referenced codes and standards.** The codes and standards referenced in this code shall be those that are listed in Chapter 8 and considered part of the requirements of this code to the prescribed extent of each such reference and as further regulated in Sections 102.7.1 and 102.7.2.

> **Exception:** Where enforcement of a code provision would violate the conditions of the listing of the equipment or appliance, the conditions of the listing shall apply.

**[A] 102.7.1 Conflicts.** Where conflicts occur between provisions of this code and the referenced standards, the provisions of this code shall apply.

**[A] 102.7.2 Provisions in referenced codes and standards.** Where the extent of the reference to a referenced code or standard includes subject matter that is within the scope of this code, the provisions of this code, as applicable, shall take precedence over the provisions in the referenced code or standard.

**[A] 102.8 Requirements not covered by code.** Requirements necessary for the strength, stability or proper operation of an existing fixture, structure or equipment, or for the public safety, health and general welfare, not specifically covered by this code, shall be determined by the *code official*.

**[A] 102.9 Application of references.** References to chapter or section numbers, or to provisions not specifically identified by number, shall be construed to refer to such chapter, section or provision of this code.

**[A] 102.10 Other laws.** The provisions of this code shall not be deemed to nullify any provisions of local, state or federal law.

## PART 2 — ADMINISTRATION AND ENFORCEMENT

*Deleted  15.22.090*

## ~~SECTION 103~~
## ~~DEPARTMENT OF PROPERTY MAINTENANCE INSPECTION~~

**[A] 103.1 General.** The department of property maintenance inspection is hereby created and the executive official in charge thereof shall be known as the *code official*.

**[A] 103.2 Appointment.** The *code official* shall be appointed by the chief appointing authority of the jurisdiction.

**[A] 103.3 Deputies.** In accordance with the prescribed procedures of this jurisdiction and with the concurrence of the appointing authority, the *code official* shall have the authority to appoint a deputy(s). Such employees shall have powers as delegated by the *code official*.

**[A] 103.4 Liability.** The *code official*, member of the board of appeals or employee charged with the enforcement of this code, while acting for the jurisdiction, in good faith and without malice in the discharge of the duties required by this code or other pertinent law or ordinance, shall not thereby be rendered liable personally, and is hereby relieved from all personal liability for any damage accruing to persons or property as a result of an act or by reason of an act or omission in the discharge of official duties. Any suit instituted against any officer or employee because of an act performed by that officer or employee in the lawful discharge of duties and under the provisions of this code shall be defended by the legal representative of the jurisdiction until the final termination of the proceedings. The *code official* or any subordinate shall not be liable for costs in an action, suit or proceeding that is instituted in pursuance of the provisions of this code.

**[A] 103.5 Fees.** The fees for activities and services performed by the department in carrying out its responsibilities under this code shall be as indicated in the following schedule.

[JURISDICTION TO INSERT APPROPRIATE SCHEDULE.]

## SECTION 104
## DUTIES AND POWERS OF THE CODE OFFICIAL

**[A] 104.1 General.** The *code official* is hereby authorized and directed to enforce the provisions of this code. The *code official* shall have the authority to render interpretations of this code and to adopt policies and procedures in order to clarify the application of its provisions. Such interpretations, policies and procedures shall be in compliance with the intent and purpose of this code. Such policies and procedures shall not have the effect of waiving requirements specifically provided for in this code.

**[A] 104.2 Inspections.** The *code official* shall make all of the required inspections, or shall accept reports of inspection by *approved* agencies or individuals. All reports of such inspections shall be in writing and be certified by a responsible officer of such *approved* agency or by the responsible individual. The *code official* is authorized to engage such expert opinion as deemed necessary to report upon unusual technical issues that arise, subject to the approval of the appointing authority.

**[A] 104.3 Right of entry.** Where it is necessary to make an inspection to enforce the provisions of this code, or whenever the *code official* has reasonable cause to believe that there exists in a *structure* or upon a *premises* a condition in violation of this code, the *code official* is authorized to enter the structure or *premises* at reasonable times to inspect or perform the duties imposed by this code, provided that if such *structure* or *premises* is occupied the *code official* shall present credentials to the *occupant* and request entry. If such structure or *premises* is unoccupied, the *code official* shall first make a reasonable effort to locate the *owner* or other person having charge or control of the *structure* or *premises* and request entry. If entry is refused, the *code official* shall have recourse to the remedies provided by law to secure entry.

**[A] 104.4 Identification.** The *code official* shall carry proper identification when inspecting *structure*s or *premises* in the performance of duties under this code.

**[A] 104.5 Notices and orders.** The *code official* shall issue all necessary notices or orders to ensure compliance with this code.

**[A] 104.6 Department records.** The *code official* shall keep official records of all business and activities of the department specified in the provisions of this code. Such records shall be retained in the official records for the period required for retention of public records.

## SECTION 105
## APPROVAL

**[A] 105.1 Modifications.** Whenever there are practical difficulties involved in carrying out the provisions of this code, the *code official* shall have the authority to grant modifications for individual cases upon application of the *owner* or *owner*'s representative, provided the *code official* shall first find that special individual reason makes the strict letter of this code impractical and the modification is in compliance with the intent and purpose of this code and that such modification does not lessen health, life and fire safety requirements. The details of action granting modifications shall be recorded and entered in the department files.

**[A] 105.2 Alternative materials, methods and equipment.** The provisions of this code are not intended to prevent the installation of any material or to prohibit any method of construction not specifically prescribed by this code, provided that any such alternative has been *approved*. An alternative material or method of construction shall be *approved* where the *code official* finds that the proposed design is satisfactory and complies with the intent of the provisions of this code, and that the material, method or work offered is, for the purpose intended, at least the equivalent of that prescribed in this code in quality, strength, effectiveness, fire resistance, durability and safety.

**[A] 105.3 Required testing.** Whenever there is insufficient evidence of compliance with the provisions of this code, or evidence that a material or method does not conform to the requirements of this code, or in order to substantiate claims for alternative materials or methods, the *code official* shall have the authority to require tests to be made as evidence of compliance at no expense to the jurisdiction.

**[A] 105.3.1 Test methods.** Test methods shall be as specified in this code or by other recognized test standards. In the absence of recognized and accepted test methods, the *code official* shall be permitted to approve appropriate testing procedures performed by an *approved* agency.

**[A] 105.3.2 Test reports.** Reports of tests shall be retained by the *code official* for the period required for retention of public records.

**[A] 105.4 Used material and equipment.** The use of used materials which meet the requirements of this code for new materials is permitted. Materials, equipment and devices shall not be reused unless such elements are in good repair or have been reconditioned and tested when necessary, placed in good and proper working condition and *approved* by the *code official*.

**[A] 105.5 Approved materials and equipment.** Materials, equipment and devices *approved* by the *code official* shall be constructed and installed in accordance with such approval.

**[A] 105.6 Research reports.** Supporting data, where necessary to assist in the approval of materials or assemblies not specifically provided for in this code, shall consist of valid research reports from *approved* sources.

## SECTION 106
## VIOLATIONS

**[A] 106.1 Unlawful acts.** ~~It shall be unlawful for a person, firm or corporation to be in conflict with or in violation of any of the provisions of this code.~~ *Amended 15.22.050*

**[A] 106.2 Notice of violation.** The *code official* shall serve a notice of violation or order in accordance with Section 107.

**[A] 106.3 Prosecution of violation.** Any person failing to comply with a notice of violation or order served in accordance with Section 107 shall be deemed guilty of a misdemeanor or civil infraction as determined by the local municipality, and the violation shall be deemed a *strict liability offense*. If the notice of violation is not complied with, the *code official* shall institute the appropriate proceeding at law or in equity to restrain, correct or abate such violation, or to require the removal or termination of the unlawful *occupancy* of the structure in violation of the provisions of this code or of the order or direction made pursuant thereto. Any action taken by the authority having jurisdiction on such *premises* shall be charged against the real estate upon which the structure is located and shall be a lien upon such real estate.

**[A] 106.4 Violation penalties.** Any person who shall violate a provision of this code, or fail to comply therewith, or with any of the requirements thereof, shall be prosecuted within the limits provided by state or local laws. Each day that a vio-

lation continues after due notice has been served shall be deemed a separate offense.

**[A] 106.5 Abatement of violation.** The imposition of the penalties herein prescribed shall not preclude the legal officer of the jurisdiction from instituting appropriate action to restrain, correct or abate a violation, or to prevent illegal *occupancy* of a building, structure or *premises*, or to stop an illegal act, conduct, business or utilization of the building, structure or *premises*.

## SECTION 107
## NOTICES AND ORDERS

**[A] 107.1 Notice to person responsible.** Whenever the *code official* determines that there has been a violation of this code or has grounds to believe that a violation has occurred, notice shall be given in the manner prescribed in Sections 107.2 and 107.3 to the person responsible for the violation as specified in this code. Notices for condemnation procedures shall also comply with Section 108.3.

**[A] 107.2 Form.** Such notice prescribed in Section 107.1 shall be in accordance with all of the following:

1. Be in writing.

2. Include a description of the real estate sufficient for identification.

3. Include a statement of the violation or violations and why the notice is being issued.

4. Include a correction order allowing a reasonable time to make the repairs and improvements required to bring the *dwelling unit* or structure into compliance with the provisions of this code.

5. Inform the property *owner* of the right to appeal.

6. Include a statement of the right to file a lien in accordance with Section 106.3.

**[A] 107.3 Method of service.** Such notice shall be deemed to be properly served if a copy thereof is:

1. Delivered personally;

2. Sent by certified or first-class mail addressed to the last known address; or

3. If the notice is returned showing that the letter was not delivered, a copy thereof shall be posted in a conspicuous place in or about the structure affected by such notice.

**[A] 107.4 Unauthorized tampering.** Signs, tags or seals posted or affixed by the *code official* shall not be mutilated, destroyed or tampered with, or removed without authorization from the *code official*. *Amended 15.22.060*

**[A] 107.5 Penalties.** ~~Penalties for noncompliance with orders and notices shall be as set forth in Section 106.4.~~

**[A] 107.6 Transfer of ownership.** It shall be unlawful for the *owner* of any *dwelling unit* or structure who has received a compliance order or upon whom a notice of violation has been served to sell, transfer, mortgage, lease or otherwise dispose of such *dwelling unit* or structure to another until the provisions of the compliance order or notice of violation have been complied with, or until such *owner* shall first furnish the grantee, transferee, mortgagee or lessee a true copy of any compliance order or notice of violation issued by the *code official* and shall furnish to the *code official* a signed and notarized statement from the grantee, transferee, mortgagee or lessee, acknowledging the receipt of such compliance order or notice of violation and fully accepting the responsibility without condition for making the corrections or repairs required by such compliance order or notice of violation.

## SECTION 108
## UNSAFE STRUCTURES AND EQUIPMENT

[A] **108.1 General.** When a structure or equipment is found by the *code official* to be unsafe, or when a structure is found unfit for human *occupancy*, or is found unlawful, such structure shall be *condemned* pursuant to the provisions of this code.

[A] **108.1.1 Unsafe structures.** An unsafe structure is one that is found to be dangerous to the life, health, property or safety of the public or the *occupants* of the structure by not providing minimum safeguards to protect or warn *occupants* in the event of fire, or because such structure contains unsafe equipment or is so damaged, decayed, dilapidated, structurally unsafe or of such faulty construction or unstable foundation, that partial or complete collapse is possible.

[A] **108.1.2 Unsafe equipment.** Unsafe equipment includes any boiler, heating equipment, elevator, moving stairway, electrical wiring or device, flammable liquid containers or other equipment on the *premises* or within the structure which is in such disrepair or condition that such equipment is a hazard to life, health, property or safety of the public or *occupants* of the *premises* or structure.

[A] **108.1.3 Structure unfit for human occupancy.** A structure is unfit for human *occupancy* whenever the *code official* finds that such structure is unsafe, unlawful or, because of the degree to which the structure is in disrepair or lacks maintenance, is insanitary, vermin or rat infested, contains filth and contamination, or lacks *ventilation*, illumination, sanitary or heating facilities or other essential equipment required by this code, or because the location of the structure constitutes a hazard to the *occupants* of the structure or to the public.

[A] **108.1.4 Unlawful structure.** An unlawful structure is one found in whole or in part to be occupied by more persons than permitted under this code, or was erected, altered or occupied contrary to law.

[A] **108.1.5 Dangerous** *structure* **or** *premises*. For the purpose of this code, any structure or *premises* that has any or all of the conditions or defects described below shall be considered dangerous:

1. Any door, aisle, passageway, stairway, exit or other means of egress that does not conform to the *approved* building or fire code of the jurisdiction as related to the requirements for existing buildings.

2. The walking surface of any aisle, passageway, stairway, exit or other means of egress is so warped, worn loose, torn or otherwise unsafe as to not provide safe and adequate means of egress.

3. Any portion of a building, structure or appurtenance that has been damaged by fire, earthquake, wind, flood, *deterioration*, *neglect*, abandonment, vandalism or by any other cause to such an extent that it is likely to partially or completely collapse, or to become *detached* or dislodged.

4. Any portion of a building, or any member, appurtenance or ornamentation on the exterior thereof that is not of sufficient strength or stability, or is not so *anchored*, attached or fastened in place so as to be capable of resisting natural or artificial loads of one and one-half the original designed value.

5. The building or structure, or part of the building or structure, because of dilapidation, *deterioration*, decay, faulty construction, the removal or movement of some portion of the ground necessary for the support, or for any other reason, is likely to partially or completely collapse, or some portion of the foundation or underpinning of the building or structure is likely to fail or give way.

6. The building or structure, or any portion thereof, is clearly unsafe for its use and *occupancy*.

7. The building or structure is *neglected*, damaged, dilapidated, unsecured or abandoned so as to become an attractive nuisance to children who might play in the building or structure to their danger, becomes a harbor for vagrants, criminals or immoral persons, or enables persons to resort to the building or structure for committing a nuisance or an unlawful act.

8. Any building or structure has been constructed, exists or is maintained in violation of any specific requirement or prohibition applicable to such building or structure provided by the *approved* building or fire code of the jurisdiction, or of any law or ordinance to such an extent as to present either a substantial risk of fire, building collapse or any other threat to life and safety.

9. A building or structure, used or intended to be used for dwelling purposes, because of inadequate maintenance, dilapidation, decay, damage, faulty construction or arrangement, inadequate light, *ventilation*, mechanical or plumbing system, or otherwise, is determined by the *code official* to be unsanitary, unfit for human habitation or in such a condition that is likely to cause sickness or disease.

10. Any building or structure, because of a lack of sufficient or proper fire-resistance-rated construction, fire protection systems, electrical system, fuel con-

nections, mechanical system, plumbing system or other cause, is determined by the *code official* to be a threat to life or health.

11. Any portion of a building remains on a site after the demolition or destruction of the building or structure or whenever any building or structure is abandoned so as to constitute such building or portion thereof as an attractive nuisance or hazard to the public.

**[A] 108.2 Closing of vacant structures.** If the structure is vacant and unfit for human habitation and *occupancy*, and is not in danger of structural collapse, the *code official* is authorized to post a placard of condemnation on the *premises* and order the structure closed up so as not to be an attractive nuisance. Upon failure of the *owner* to close up the *premises* within the time specified in the order, the *code official* shall cause the *premises* to be closed and secured through any available public agency or by contract or arrangement by private persons and the cost thereof shall be charged against the real estate upon which the structure is located and shall be a lien upon such real estate and may be collected by any other legal resource.

**[A] 108.2.1 Authority to disconnect service utilities.** The *code official* shall have the authority to authorize disconnection of utility service to the building, structure or system regulated by this code and the referenced codes and standards set forth in Section 102.7 in case of emergency where necessary to eliminate an immediate hazard to life or property or when such utility connection has been made without approval. The *code official* shall notify the serving utility and, whenever possible, the *owner* and *occupant* of the building, structure or service system of the decision to disconnect prior to taking such action. If not notified prior to disconnection the *owner* or *occupant* of the building structure or service system shall be notified in writing as soon as practical thereafter.

**[A] 108.3 Notice.** Whenever the *code official* has *condemned* a structure or equipment under the provisions of this section, notice shall be posted in a conspicuous place in or about the structure affected by such notice and served on the *owner* or the person or persons responsible for the structure or equipment in accordance with Section 107.3. If the notice pertains to equipment, it shall also be placed on the *condemned* equipment. The notice shall be in the form prescribed in Section 107.2.

**[A] 108.4 Placarding.** Upon failure of the *owner* or person responsible to comply with the notice provisions within the time given, the *code official* shall post on the *premises* or on defective equipment a placard bearing the word "Condemned" and a statement of the penalties provided for occupying the *premises*, operating the equipment or removing the placard.

**[A] 108.4.1 Placard removal.** The *code official* shall remove the condemnation placard whenever the defect or defects upon which the condemnation and placarding action were based have been eliminated. Any person who

defaces or removes a condemnation placard without the approval of the *code official* shall be subject to the penalties provided by this code.

**[A] 108.5 Prohibited occupancy.** Any occupied structure *condemned* and placarded by the *code official* shall be vacated as ordered by the *code official*. Any person who shall occupy a placarded *premises* or shall operate placarded equipment, and any *owner* or any person responsible for the *premises* who shall let anyone occupy a placarded *premises* or operate placarded equipment shall be liable for the penalties provided by this code.

**[A] 108.6 Abatement methods.** The *owner*, *operator* or *occupant* of a building, *premises* or equipment deemed unsafe by the *code official* shall abate or cause to be abated or corrected such unsafe conditions either by repair, rehabilitation, demolition or other *approved* corrective action.

**[A] 108.7 Record.** The *code official* shall cause a report to be filed on an unsafe condition. The report shall state the *occupancy* of the structure and the nature of the unsafe condition.

## SECTION 109
## EMERGENCY MEASURES

**[A] 109.1 Imminent danger.** When, in the opinion of the *code official*, there is *imminent danger* of failure or collapse of a building or structure which endangers life, or when any structure or part of a structure has fallen and life is endangered by the occupation of the structure, or when there is actual or potential danger to the building *occupants* or those in the proximity of any structure because of explosives, explosive fumes or vapors or the presence of toxic fumes, gases or materials, or operation of defective or dangerous equipment, the *code official* is hereby authorized and empowered to order and require the *occupants* to vacate the *premises* forthwith. The *code official* shall cause to be posted at each entrance to such structure a notice reading as follows: "This *Structure* Is Unsafe and Its *Occupancy* Has Been Prohibited by the *Code Official*." It shall be unlawful for any person to enter such structure except for the purpose of securing the structure, making the required repairs, removing the hazardous condition or of demolishing the same.

**[A] 109.2 Temporary safeguards.** Notwithstanding other provisions of this code, whenever, in the opinion of the *code official*, there is *imminent danger* due to an unsafe condition, the *code official* shall order the necessary work to be done, including the boarding up of openings, to render such structure temporarily safe whether or not the legal procedure herein described has been instituted; and shall cause such other action to be taken as the *code official* deems necessary to meet such emergency.

**[A] 109.3 Closing streets.** When necessary for public safety, the *code official* shall temporarily close structures and close, or order the authority having jurisdiction to close, sidewalks, streets, *public ways* and places adjacent to unsafe structures, and prohibit the same from being utilized.

**[A] 109.4 Emergency repairs.** For the purposes of this section, the *code official* shall employ the necessary labor and materials to perform the required work as expeditiously as possible.

**[A] 109.5 Costs of emergency repairs.** Costs incurred in the performance of emergency work shall be paid by the jurisdiction. The legal counsel of the jurisdiction shall institute appropriate action against the *owner* of the *premises* where the unsafe structure is or was located for the recovery of such costs.

**[A] 109.6 Hearing.** Any person ordered to take emergency measures shall comply with such order forthwith. Any affected person shall thereafter, upon petition directed to the appeals board, be afforded a hearing as described in this code.

## SECTION 110
## DEMOLITION

**[A] 110.1 General.** The *code official* shall order the *owner* of any *premises* upon which is located any structure, which in the *code official* judgment after review is so deteriorated or dilapidated or has become so out of repair as to be dangerous, unsafe, insanitary or otherwise unfit for human habitation or occupancy, and such that it is unreasonable to repair the structure, to demolish and remove such structure; or if such structure is capable of being made safe by repairs, to repair and make safe and sanitary, or to board up and hold for future repair or to demolish and remove at the *owner's* option; or where there has been a cessation of normal construction of any structure for a period of more than two years, the *code official* shall order the *owner* to demolish and remove such structure, or board up until future repair. Boarding the building up for future repair shall not extend beyond one year, unless *approved* by the building official.

**[A] 110.2 Notices and orders.** All notices and orders shall comply with Section 107.

**[A] 110.3 Failure to comply.** If the *owner* of a *premises* fails to comply with a demolition order within the time prescribed, the *code official* shall cause the structure to be demolished and removed, either through an available public agency or by contract or arrangement with private persons, and the cost of such demolition and removal shall be charged against the real estate upon which the structure is located and shall be a lien upon such real estate.

**[A] 110.4 Salvage materials.** When any structure has been ordered demolished and removed, the governing body or other designated officer under said contract or arrangement aforesaid shall have the right to sell the salvage and valuable materials at the highest price obtainable. The net proceeds of such sale, after deducting the expenses of such demolition and removal, shall be promptly remitted with a report of such sale or transaction, including the items of expense and the amounts deducted, for the person who is entitled thereto, subject to any order of a court. If such a surplus does not remain to be turned over, the report shall so state.

## SECTION 111 Amended
## MEANS OF APPEAL 15-22070

**[A] 111.1 Application for appeal.** Any person directly affected by a decision of the *code official* or a notice or order issued under this code shall have the right to appeal to the board of appeals, provided that a written application for appeal is filed within 20 days after the day the decision, notice or order was served. An application for appeal shall be based on a claim that the true intent of this code or the rules legally adopted thereunder have been incorrectly interpreted, the provisions of this code do not fully apply, or the requirements of this code are adequately satisfied by other means.

**[A] 111.2 Membership of board.** The board of appeals shall consist of a minimum of three members who are qualified by experience and training to pass on matters pertaining to property maintenance and who are not employees of the jurisdiction. The *code official* shall be an ex-officio member but shall have no vote on any matter before the board. The board shall be appointed by the chief appointing authority, and shall serve staggered and overlapping terms.

**[A] 111.2.1 Alternate members.** The chief appointing authority shall appoint a minimum of two alternate members who shall be called by the board chairman to hear appeals during the absence or disqualification of a member. Alternate members shall possess the qualifications required for board membership.

**[A] 111.2.2 Chairman.** The board shall annually select one of its members to serve as chairman.

**[A] 111.2.3 Disqualification of member.** A member shall not hear an appeal in which that member has a personal, professional or financial interest.

**[A] 111.2.4 Secretary.** The chief administrative officer shall designate a qualified person to serve as secretary to the board. The secretary shall file a detailed record of all proceedings in the office of the chief administrative officer.

**[A] 111.2.5 Compensation of members.** Compensation of members shall be determined by law.

**[A] 111.3 Notice of meeting.** The board shall meet upon notice from the chairman, within 20 days of the filing of an appeal, or at stated periodic meetings.

**[A] 111.4 Open hearing.** All hearings before the board shall be open to the public. The appellant, the appellant's representative, the *code official* and any person whose interests are affected shall be given an opportunity to be heard. A quorum shall consist of a minimum of two-thirds of the board membership.

**[A] 111.4.1 Procedure.** The board shall adopt and make available to the public through the secretary procedures under which a hearing will be conducted. The procedures shall not require compliance with strict rules of evidence, but shall mandate that only relevant information be received.

**[A] 111.5 Postponed hearing.** When the full board is not present to hear an appeal, either the appellant or the appellant's representative shall have the right to request a postponement of the hearing.

**[A] 111.6 Board decision.** The board shall modify or reverse the decision of the *code official* only by a concurring vote of a majority of the total number of appointed board members.

**[A] 111.6.1 Records and copies.** The decision of the board shall be recorded. Copies shall be furnished to the appellant and to the *code official*.

**[A] 111.6.2 Administration.** The *code official* shall take immediate action in accordance with the decision of the board.

**[A] 111.7 Court review.** Any person, whether or not a previous party of the appeal, shall have the right to apply to the appropriate court for a writ of certiorari to correct errors of law. Application for review shall be made in the manner and time required by law following the filing of the decision in the office of the chief administrative officer.

**[A] 111.8 Stays of enforcement.** Appeals of notice and orders (other than *Imminent Danger* notices) shall stay the enforcement of the notice and order until the appeal is heard by the appeals board.

## SECTION 112
## STOP WORK ORDER

**[A] 112.1 Authority.** Whenever the *code official* finds any work regulated by this code being performed in a manner contrary to the provisions of this code or in a dangerous or unsafe manner, the *code official* is authorized to issue a stop work order.

**[A] 112.2 Issuance.** A stop work order shall be in writing and shall be given to the *owner* of the property, to the *owner's* agent, or to the person doing the work. Upon issuance of a stop work order, the cited work shall immediately cease. The stop work order shall state the reason for the order and the conditions under which the cited work is authorized to resume.

**[A] 112.3 Emergencies.** Where an emergency exists, the *code official* shall not be required to give a written notice prior to stopping the work.

**[A] 112.4 Failure to comply.** Any person who shall continue any work after having been served with a stop work order, except such work as that person is directed to perform to remove a violation or unsafe condition, shall be liable to a fine of not less than [AMOUNT] dollars or more than [AMOUNT] dollars.

# CHAPTER 2
# DEFINITIONS

## SECTION 201
## GENERAL

**201.1 Scope.** Unless otherwise expressly stated, the following terms shall, for the purposes of this code, have the meanings shown in this chapter.

**201.2 Interchangeability.** Words stated in the present tense include the future; words stated in the masculine gender include the feminine and neuter; the singular number includes the plural and the plural, the singular.

**201.3 Terms defined in other codes.** Where terms are not defined in this code and are defined in the *International Building Code, International Existing Building Code, International Fire Code, International Fuel Gas Code, International Mechanical Code, International Plumbing Code, International Residential Code, International Zoning Code* or NFPA 70, such terms shall have the meanings ascribed to them as stated in those codes.

**201.4 Terms not defined.** Where terms are not defined through the methods authorized by this section, such terms shall have ordinarily accepted meanings such as the context implies.

**201.5 Parts.** Whenever the words "*dwelling unit,*" "*dwelling,*" "*premises,*" "building," "*rooming house,*" "*rooming unit,*" "*housekeeping unit*" or "story" are stated in this code, they shall be construed as though they were followed by the words "or any part thereof."

## SECTION 202
## GENERAL DEFINITIONS

**ANCHORED.** Secured in a manner that provides positive connection.

**[A] APPROVED.** *Approved* by the *code official.*

**BASEMENT.** That portion of a building which is partly or completely below grade.

**BATHROOM.** A room containing plumbing fixtures including a bathtub or shower.

**BEDROOM.** Any room or space used or intended to be used for sleeping purposes in either a dwelling or *sleeping unit.*

**[A] CODE OFFICIAL.** The official who is charged with the administration and enforcement of this code, or any duly authorized representative.

**CONDEMN.** To adjudge unfit for *occupancy.*

**DETACHED.** When a structural element is physically disconnected from another and that connection is necessary to provide a positive connection.

**DETERIORATION.** To weaken, disintegrate, corrode, rust or decay and lose effectiveness.

**[B] DWELLING UNIT.** A single unit providing complete, independent living facilities for one or more persons, including permanent provisions for living, sleeping, eating, cooking and sanitation.

**[Z] EASEMENT.** That portion of land or property reserved for present or future use by a person or agency other than the legal fee *owner*(s) of the property. The *easement* shall be permitted to be for use under, on or above a said lot or lots.

**EQUIPMENT SUPPORT.** Those structural members or assemblies of members or manufactured elements, including braces, frames, lugs, snuggers, hangers or saddles, that transmit gravity load, lateral load and operating load between the equipment and the structure.

**EXTERIOR PROPERTY.** The open space on the *premises* and on adjoining property under the control of *owners* or *operators* of such *premises.*

**GARBAGE.** The animal or vegetable waste resulting from the handling, preparation, cooking and consumption of food.

**[B] GUARD.** A building component or a system of building components located at or near the open sides of elevated walking surfaces that minimizes the possibility of a fall from the walking surface to a lower level.

**[B] HABITABLE SPACE.** Space in a structure for living, sleeping, eating or cooking. *Bathrooms, toilet rooms,* closets, halls, storage or utility spaces, and similar areas are not considered *habitable spaces.*

**HOUSEKEEPING UNIT.** A room or group of rooms forming a single *habitable space* equipped and intended to be used for living, sleeping, cooking and eating which does not contain, within such a unit, a toilet, lavatory and bathtub or shower.

**IMMINENT DANGER.** A condition which could cause serious or life-threatening injury or death at any time.

**INFESTATION.** The presence, within or contiguous to, a structure or *premises* of insects, rats, vermin or other pests.

**INOPERABLE MOTOR VEHICLE.** A vehicle which cannot be driven upon the public streets for reason including but not limited to being unlicensed, wrecked, abandoned, in a state of disrepair, or incapable of being moved under its own power.

**[A] LABELED.** Equipment, materials or products to which have been affixed a label, seal, symbol or other identifying mark of a nationally recognized testing laboratory, inspection agency or other organization concerned with product evaluation that maintains periodic inspection of the production of the above-*labeled* items and whose labeling indicates either that the equipment, material or product meets identified standards or has been tested and found suitable for a specified purpose.

**LET FOR OCCUPANCY or LET.** To permit, provide or offer possession or *occupancy* of a dwelling, *dwelling unit*, *rooming unit*, building, premise or structure by a person who is or is not the legal *owner* of record thereof, pursuant to a written or unwritten lease, agreement or license, or pursuant to a recorded or unrecorded agreement of contract for the sale of land.

**NEGLECT.** The lack of proper maintenance for a building or *structure*.

**[A] OCCUPANCY.** The purpose for which a building or portion thereof is utilized or occupied.

**OCCUPANT.** Any individual living or sleeping in a building, or having possession of a space within a building.

**OPENABLE AREA.** That part of a window, skylight or door which is available for unobstructed *ventilation* and which opens directly to the outdoors.

**OPERATOR.** Any person who has charge, care or control of a structure or *premises* which is let or offered for *occupancy*.

**[A] OWNER.** Any person, agent, *operator*, firm or corporation having a legal or equitable interest in the property; or recorded in the official records of the state, county or municipality as holding title to the property; or otherwise having control of the property, including the guardian of the estate of any such person, and the executor or administrator of the estate of such person if ordered to take possession of real property by a court.

**PERSON.** An individual, corporation, partnership or any other group acting as a unit.

**PEST ELIMINATION.** The control and elimination of insects, rodents or other pests by eliminating their harborage places; by removing or making inaccessible materials that serve as their food or water; by other *approved pest elimination* methods.

**[A] PREMISES.** A lot, plot or parcel of land, *easement* or *public way*, including any structures thereon.

**[A] PUBLIC WAY.** Any street, alley or similar parcel of land essentially unobstructed from the ground to the sky, which is deeded, dedicated or otherwise permanently appropriated to the public for public use.

**ROOMING HOUSE.** A building arranged or occupied for lodging, with or without meals, for compensation and not occupied as a one- or two-family dwelling.

**ROOMING UNIT.** Any room or group of rooms forming a single habitable unit occupied or intended to be occupied for sleeping or living, but not for cooking purposes.

**RUBBISH.** Combustible and noncombustible waste materials, except garbage; the term shall include the residue from the burning of wood, coal, coke and other combustible materials, paper, rags, cartons, boxes, wood, excelsior, rubber, leather, tree branches, *yard* trimmings, tin cans, metals, mineral matter, glass, crockery and dust and other similar materials.

**[B] SLEEPING UNIT.** A room or space in which people sleep, which can also include permanent provisions for living, eating and either sanitation or kitchen facilities, but not both. Such rooms and spaces that are also part of a *dwelling unit* are not *sleeping units*.

**STRICT LIABILITY OFFENSE.** An offense in which the prosecution in a legal proceeding is not required to prove criminal intent as a part of its case. It is enough to prove that the defendant either did an act which was prohibited, or failed to do an act which the defendant was legally required to do.

**[A] STRUCTURE.** That which is built or constructed or a portion thereof.

**TENANT.** A person, corporation, partnership or group, whether or not the legal *owner* of record, occupying a building or portion thereof as a unit.

**TOILET ROOM.** A room containing a water closet or urinal but not a bathtub or shower.

**ULTIMATE DEFORMATION.** The deformation at which failure occurs and which shall be deemed to occur if the sustainable load reduces to 80 percent or less of the maximum strength.

**[M] VENTILATION.** The natural or mechanical process of supplying conditioned or unconditioned air to, or removing such air from, any space.

**WORKMANLIKE.** Executed in a skilled manner; e.g., generally plumb, level, square, in line, undamaged and without marring adjacent work.

**[Z] YARD.** An open space on the same lot with a structure.

# CHAPTER 3

# GENERAL REQUIREMENTS

## SECTION 301
## GENERAL

**301.1 Scope.** The provisions of this chapter shall govern the minimum conditions and the responsibilities of persons for maintenance of structures, equipment and *exterior property*.

**301.2 Responsibility.** The *owner* of the *premises* shall maintain the structures and *exterior property* in compliance with these requirements, except as otherwise provided for in this code. A person shall not occupy as owner-occupant or permit another person to occupy *premises* which are not in a sanitary and safe condition and which do not comply with the requirements of this chapter. *Occupants* of a *dwelling unit*, *rooming unit* or *housekeeping unit* are responsible for keeping in a clean, sanitary and safe condition that part of the *dwelling unit*, *rooming unit*, *housekeeping unit* or *premises* which they occupy and control.

**301.3 Vacant structures and land.** All vacant structures and *premises* thereof or vacant land shall be maintained in a clean, safe, secure and sanitary condition as provided herein so as not to cause a blighting problem or adversely affect the public health or safety.

## SECTION 302
## EXTERIOR PROPERTY AREAS

*Deleted 15.22.080*

**302.1 Sanitation.** All *exterior property* and *premises* shall be maintained in a clean, safe and sanitary condition. The *occupant* shall keep that part of the *exterior property* which such *occupant* occupies or controls in a clean and sanitary condition.

**302.2 Grading and drainage.** All *premises* shall be graded and maintained to prevent the erosion of soil and to prevent the accumulation of stagnant water thereon, or within any structure located thereon.

**Exception:** *Approved* retention areas and reservoirs.

**302.3 Sidewalks and driveways.** All sidewalks, walkways, stairs, driveways, parking spaces and similar areas shall be kept in a proper state of repair, and maintained free from hazardous conditions.

**302.4 Weeds.** All *premises* and *exterior property* shall be maintained free from weeds or plant growth in excess of [JURISDICTION TO INSERT HEIGHT IN INCHES]. All noxious weeds shall be prohibited. Weeds shall be defined as all grasses, annual plants and vegetation, other than trees or shrubs provided; however, this term shall not include cultivated flowers and gardens.

Upon failure of the *owner* or agent having charge of a property to cut and destroy weeds after service of a notice of violation, they shall be subject to prosecution in accordance with Section 106.3 and as prescribed by the authority having jurisdiction. Upon failure to comply with the notice of viola-

tion, any duly authorized employee of the jurisdiction or contractor hired by the jurisdiction shall be authorized to enter upon the property in violation and cut and destroy the weeds growing thereon, and the costs of such removal shall be paid by the *owner* or agent responsible for the property.

**302.5 Rodent harborage.** All structures and *exterior property* shall be kept free from rodent harborage and *infestation*. Where rodents are found, they shall be promptly exterminated by *approved* processes which will not be injurious to human health. After pest elimination, proper precautions shall be taken to eliminate rodent harborage and prevent reinfestation.

**302.6 Exhaust vents.** Pipes, ducts, conductors, fans or blowers shall not discharge gases, steam, vapor, hot air, grease, smoke, odors or other gaseous or particulate wastes directly upon abutting or adjacent public or private property or that of another *tenant*.

**302.7 Accessory structures.** All accessory structures, including *detached* garages, fences and walls, shall be maintained structurally sound and in good repair.

**302.8 Motor vehicles.** Except as provided for in other regulations, no inoperative or unlicensed motor vehicle shall be parked, kept or stored on any *premises*, and no vehicle shall at any time be in a state of major disassembly, disrepair, or in the process of being stripped or dismantled. Painting of vehicles is prohibited unless conducted inside an *approved* spray booth.

**Exception:** A vehicle of any type is permitted to undergo major overhaul, including body work, provided that such work is performed inside a structure or similarly enclosed area designed and *approved* for such purposes.

**302.9 Defacement of property.** No person shall willfully or wantonly damage, mutilate or deface any exterior surface of any structure or building on any private or public property by placing thereon any marking, carving or graffiti.

It shall be the responsibility of the *owner* to restore said surface to an *approved* state of maintenance and repair.

*Deleted 15.22.090*

## SECTION 303
## SWIMMING POOLS, SPAS AND HOT TUBS

**303.1 Swimming pools.** Swimming pools shall be maintained in a clean and sanitary condition, and in good repair.

**303.2 Enclosures.** Private swimming pools, hot tubs and spas, containing water more than 24 inches (610 mm) in depth shall be completely surrounded by a fence or barrier at least 48 inches (1219 mm) in height above the finished ground level measured on the side of the barrier away from the pool. Gates and doors in such barriers shall be self-closing and self-latching. Where the self-latching device is a minimum of 54 inches (1372 mm) above the bottom of the gate, the release mechanism shall be located on the pool side of the gate. Self-closing and self-latching gates shall be maintained such that the gate will positively close and latch when released from an open position of 6 inches (152 mm) from the gatepost. No existing pool enclosure shall be removed, replaced or changed in a manner that reduces its effectiveness as a safety barrier.

**Exception:** Spas or hot tubs with a safety cover that complies with ASTM F 1346 shall be exempt from the provisions of this section.

## SECTION 304
## EXTERIOR STRUCTURE  ~~Deleted 15.22.100~~

**304.1 General.** The exterior of a structure shall be maintained in good repair, structurally sound and sanitary so as not to pose a threat to the public health, safety or welfare.

**304.1.1 Unsafe conditions.** The following conditions shall be determined as unsafe and shall be repaired or replaced to comply with the *International Building Code* or the *International Existing Building Code* as required for existing buildings:

1. The nominal strength of any structural member is exceeded by nominal loads, the load effects or the required strength;

2. The *anchorage* of the floor or roof to walls or columns, and of walls and columns to foundations is not capable of resisting all nominal loads or load effects;

3. Structures or components thereof that have reached their limit state;

4. Siding and masonry joints including joints between the building envelope and the perimeter of windows, doors and skylights are not maintained, weather resistant or water tight;

5. Structural members that have evidence of *deterioration* or that are not capable of safely supporting all nominal loads and load effects;

6. Foundation systems that are not firmly supported by footings, are not plumb and free from open cracks and breaks, are not properly *anchored* or are not capable of supporting all nominal loads and resisting all load effects;

7. Exterior walls that are not *anchored* to supporting and supported elements or are not plumb and free of holes, cracks or breaks and loose or rotting materials, are not properly *anchored* or are not capable of supporting all nominal loads and resisting all load effects;

8. Roofing or roofing components that have defects that admit rain, roof surfaces with inadequate drainage, or any portion of the roof framing that is not in good repair with signs of *deterioration*, fatigue or without proper anchorage and incapable of supporting all nominal loads and resisting all load effects;

9. Flooring and flooring components with defects that affect serviceability or flooring components that show signs of *deterioration* or fatigue, are not properly *anchored* or are incapable of supporting all nominal loads and resisting all load effects;

10. Veneer, cornices, belt courses, corbels, trim, wall facings and similar decorative features not properly anchored or that are anchored with connections not capable of supporting all nominal loads and resisting all load effects;

11. Overhang extensions or projections including, but not limited to, trash chutes, canopies, marquees, signs, awnings, fire escapes, standpipes and exhaust ducts not properly *anchored* or that are *anchored* with connections not capable of supporting all nominal loads and resisting all load effects;

12. Exterior stairs, decks, porches, balconies and all similar appurtenances attached thereto, including *guard*s and handrails, are not structurally sound, not properly *anchored* or that are *anchored* with connections not capable of supporting all nominal loads and resisting all load effects; or

13. Chimneys, cooling towers, smokestacks and similar appurtenances not structurally sound or not properly *anchored,* or that are anchored with connections not capable of supporting all nominal loads and resisting all load effects.

**Exceptions:**

1. When substantiated otherwise by an *approved* method.

2. Demolition of unsafe conditions shall be permitted when *approved* by the *code official.*

**304.2 Protective treatment.** All exterior surfaces, including but not limited to, doors, door and window frames, cornices, porches, trim, balconies, decks and fences, shall be maintained in good condition. Exterior wood surfaces, other than decay-resistant woods, shall be protected from the elements and decay by painting or other protective covering or treatment. Peeling, flaking and chipped paint shall be eliminated and surfaces repainted. All siding and masonry joints, as well as those between the building envelope and the perimeter of windows, doors and skylights, shall be maintained weather resistant and water tight. All metal surfaces subject to rust or corrosion shall be coated to inhibit such rust and corrosion, and all surfaces with rust or corrosion shall be stabilized and coated to inhibit future rust and corrosion. Oxidation stains shall be removed from exterior surfaces. Surfaces designed for stabilization by oxidation are exempt from this requirement.

**[F] 304.3 Premises identification.** Buildings shall have *approved* address numbers placed in a position to be plainly legible and visible from the street or road fronting the property. These numbers shall contrast with their background. Address numbers shall be Arabic numerals or alphabet letters. Numbers shall be a minimum of 4 inches (102 mm) in height with a minimum stroke width of 0.5 inch (12.7 mm).

**304.4 Structural members.** All structural members shall be maintained free from *deterioration*, and shall be capable of safely supporting the imposed dead and live loads.

**304.5 Foundation walls.** All foundation walls shall be maintained plumb and free from open cracks and breaks and shall be kept in such condition so as to prevent the entry of rodents and other pests.

**304.6 Exterior walls.** All exterior walls shall be free from holes, breaks, and loose or rotting materials; and maintained weatherproof and properly surface coated where required to prevent *deterioration*.

**304.7 Roofs and drainage.** The roof and flashing shall be sound, tight and not have defects that admit rain. Roof drainage shall be adequate to prevent dampness or *deterioration* in the walls or interior portion of the structure. Roof drains, gutters and downspouts shall be maintained in good repair and free from obstructions. Roof water shall not be discharged in a manner that creates a public nuisance.

**304.8 Decorative features.** All cornices, belt courses, corbels, terra cotta trim, wall facings and similar decorative features shall be maintained in good repair with proper anchorage and in a safe condition.

**304.9 Overhang extensions.** All overhang extensions including, but not limited to canopies, marquees, signs, metal awnings, fire escapes, standpipes and exhaust ducts shall be maintained in good repair and be properly *anchored* so as to be kept in a sound condition. When required, all exposed surfaces of metal or wood shall be protected from the elements and against decay or rust by periodic application of weather-coating materials, such as paint or similar surface treatment.

**304.10 Stairways, decks, porches and balconies.** Every exterior stairway, deck, porch and balcony, and all appurtenances attached thereto, shall be maintained structurally sound, in good repair, with proper anchorage and capable of supporting the imposed loads.

**304.11 Chimneys and towers.** All chimneys, cooling towers, smoke stacks, and similar appurtenances shall be maintained structurally safe and sound, and in good repair. All exposed surfaces of metal or wood shall be protected from the elements and against decay or rust by periodic application of weather-coating materials, such as paint or similar surface treatment.

**304.12 Handrails and guards.** Every handrail and *guard* shall be firmly fastened and capable of supporting normally imposed loads and shall be maintained in good condition.

**304.13 Window, skylight and door frames.** Every window, skylight, door and frame shall be kept in sound condition, good repair and weather tight.

**304.13.1 Glazing.** All glazing materials shall be maintained free from cracks and holes.

**304.13.2 Openable windows.** Every window, other than a fixed window, shall be easily openable and capable of being held in position by window hardware.

**304.14 Insect screens.** During the period from [DATE] to [DATE], every door, window and other outside opening required for *ventilation* of habitable rooms, food preparation areas, food service areas or any areas where products to be included or utilized in food for human consumption are processed, manufactured, packaged or stored shall be supplied with *approved* tightly fitting screens of minimum 16 mesh per inch (16 mesh per 25 mm), and every screen door used for insect control shall have a self-closing device in good working condition.

> **Exception:** Screens shall not be required where other *approved* means, such as air curtains or insect repellent fans, are employed.

**304.15 Doors.** All exterior doors, door assemblies, operator systems if provided, and hardware shall be maintained in good condition. Locks at all entrances to dwelling units and sleeping units shall tightly secure the door. Locks on means of egress doors shall be in accordance with Section 702.3.

**304.16 Basement hatchways.** Every *basement* hatchway shall be maintained to prevent the entrance of rodents, rain and surface drainage water.

**304.17 Guards for basement windows.** Every *basement* window that is openable shall be supplied with rodent shields, storm windows or other *approved* protection against the entry of rodents.

**304.18 Building security.** Doors, windows or hatchways for *dwelling units*, room units or *housekeeping units* shall be provided with devices designed to provide security for the *occupant*s and property within.

**304.18.1 Doors.** Doors providing access to a *dwelling unit*, *rooming unit* or *housekeeping unit* that is rented, leased or let shall be equipped with a deadbolt lock designed to be readily openable from the side from which egress is to be made without the need for keys, special knowledge or effort and shall have a minimum lock throw of 1 inch (25 mm). Such deadbolt locks shall be installed according to the manufacturer's specifications and maintained in good working order. For the purpose of this section, a sliding bolt shall not be considered an acceptable deadbolt lock.

**304.18.2 Windows.** Operable windows located in whole or in part within 6 feet (1828 mm) above ground level or a walking surface below that provide access to a *dwelling unit*, *rooming unit* or *housekeeping unit* that is rented, leased or let shall be equipped with a window sash locking device.

**304.18.3 Basement hatchways.** *Basement* hatchways that provide access to a *dwelling unit*, *rooming unit* or *housekeeping unit* that is rented, leased or let shall be equipped with devices that secure the units from unauthorized entry.

**304.19 Gates.** All exterior gates, gate assemblies, operator systems if provided, and hardware shall be maintained in good condition. Latches at all entrances shall tightly secure the gates.

## SECTION 305
## INTERIOR STRUCTURE

**305.1 General.** The interior of a structure and equipment therein shall be maintained in good repair, structurally sound and in a sanitary condition. *Occupants* shall keep that part of the structure which they occupy or control in a clean and sanitary condition. Every *owner* of a structure containing a *rooming house*, *housekeeping units*, a hotel, a dormitory, two or more *dwelling units* or two or more nonresidential occupancies, shall maintain, in a clean and sanitary condition, the shared or public areas of the structure and *exterior property*.

**305.1.1 Unsafe conditions.** The following conditions shall be determined as unsafe and shall be repaired or replaced to comply with the *International Building Code* or the *International Existing Building Code* as required for existing buildings:

1. The nominal strength of any structural member is exceeded by nominal loads, the load effects or the required strength;

2. The anchorage of the floor or roof to walls or columns, and of walls and columns to foundations is not capable of resisting all nominal loads or load effects;

3. Structures or components thereof that have reached their limit state;

4. Structural members are incapable of supporting nominal loads and load effects;

5. Stairs, landings, balconies and all similar walking surfaces, including *guards* and handrails, are not structurally sound, not properly *anchored* or are *anchored* with connections not capable of supporting all nominal loads and resisting all load effects;

6. Foundation systems that are not firmly supported by footings are not plumb and free from open cracks and breaks, are not properly *anchored* or are not capable of supporting all nominal loads and resisting all load effects.

**Exceptions:**

1. When substantiated otherwise by an *approved* method.

2. Demolition of unsafe conditions shall be permitted when *approved* by the *code official*.

**305.2 Structural members.** All structural members shall be maintained structurally sound, and be capable of supporting the imposed loads.

**305.3 Interior surfaces.** All interior surfaces, including windows and doors, shall be maintained in good, clean and sanitary condition. Peeling, chipping, flaking or abraded paint shall be repaired, removed or covered. Cracked or loose plaster, decayed wood and other defective surface conditions shall be corrected.

**305.4 Stairs and walking surfaces.** Every stair, ramp, landing, balcony, porch, deck or other walking surface shall be maintained in sound condition and good repair.

**305.5 Handrails and guards.** Every handrail and *guard* shall be firmly fastened and capable of supporting normally imposed loads and shall be maintained in good condition.

**305.6 Interior doors.** Every interior door shall fit reasonably well within its frame and shall be capable of being opened and closed by being properly and securely attached to jambs, headers or tracks as intended by the manufacturer of the attachment hardware.

*Deleted 15.22.110*

## SECTION 306
## COMPONENT SERVICEABILITY

**306.1 General.** The components of a structure and equipment therein shall be maintained in good repair, structurally sound and in a sanitary condition.

**306.1.1 Unsafe conditions.** Where any of the following conditions cause the component or system to be beyond its limit state, the component or system shall be determined as unsafe and shall be repaired or replaced to comply with the *International Building Code* as required for existing buildings:

1. Soils that have been subjected to any of the following conditions:

   1.1. Collapse of footing or foundation system;

   1.2. Damage to footing, foundation, concrete or other structural element due to soil expansion;

   1.3. Adverse effects to the design strength of footing, foundation, concrete or other structural element due to a chemical reaction from the soil;

   1.4. Inadequate soil as determined by a geotechnical investigation;

   1.5. Where the allowable bearing capacity of the soil is in doubt; or

   1.6. Adverse effects to the footing, foundation, concrete or other structural element due to the ground water table.

2. Concrete that has been subjected to any of the following conditions:

   2.1. *Deterioration*;

   2.2. *Ultimate deformation*;

   2.3. Fractures;

   2.4. Fissures;

   2.5. Spalling;

   2.6. Exposed reinforcement; or

   2.7. *Detached*, dislodged or failing connections.

3. Aluminum that has been subjected to any of the following conditions:

   3.1. *Deterioration*;

   3.2. Corrosion;

   3.3. Elastic deformation;

   3.4. *Ultimate deformation*;

   3.5. Stress or strain cracks;

   3.6. Joint fatigue; or

   3.7. *Detached*, dislodged or failing connections.

4. Masonry that has been subjected to any of the following conditions:

    4.1. *Deterioration;*

    4.2. *Ultimate deformation;*

    4.3. Fractures in masonry or mortar joints;

    4.4. Fissures in masonry or mortar joints;

    4.5. Spalling;

    4.6. Exposed reinforcement; or

    4.7. *Detached,* dislodged or failing connections.

5. Steel that has been subjected to any of the following conditions:

    5.1. *Deterioration;*

    5.2. Elastic deformation;

    5.3. *Ultimate deformation;*

    5.4. Metal fatigue; or

    5.5. *Detached,* dislodged or failing connections.

6. Wood that has been subjected to any of the following conditions:

    6.1. *Ultimate deformation;*

    6.2. *Deterioration;*

    6.3. Damage from insects, rodents and other vermin;

    6.4. Fire damage beyond charring;

    6.5. Significant splits and checks;

    6.6. Horizontal shear cracks;

    6.7. Vertical shear cracks;

    6.8. Inadequate support;

    6.9. *Detached,* dislodged or failing connections; or

    6.10. Excessive cutting and notching.

**Exceptions:**

1. When substantiated otherwise by an *approved* method.

2. Demolition of unsafe conditions shall be permitted when *approved* by the *code official.*

## SECTION 307
## HANDRAILS AND GUARDRAILS

**307.1 General.** Every exterior and interior flight of stairs having more than four risers shall have a handrail on one side of the stair and every open portion of a stair, landing, balcony, porch, deck, ramp or other walking surface which is more than 30 inches (762 mm) above the floor or grade below shall have *guards.* Handrails shall not be less than 30 inches (762 mm) in height or more than 42 inches (1067 mm) in height measured vertically above the nosing of the tread or above the finished floor of the landing or walking surfaces. *Guards* shall not be less than 30 inches (762 mm) in height above the floor of the landing, balcony, porch, deck, or ramp or other walking surface.

**Exception:** *Guards* shall not be required where exempted by the adopted building code.

*Deleted*
## SECTION 308   *15.22.120*
## RUBBISH AND GARBAGE

**308.1 Accumulation of rubbish or garbage.** All *exterior property* and *premises,* and the interior of every structure, shall be free from any accumulation of *rubbish* or garbage.

**308.2 Disposal of rubbish.** Every *occupant* of a structure shall dispose of all *rubbish* in a clean and sanitary manner by placing such *rubbish* in *approved* containers.

    **308.2.1 Rubbish storage facilities.** The *owner* of every occupied *premises* shall supply *approved* covered containers for *rubbish,* and the *owner* of the *premises* shall be responsible for the removal of *rubbish.*

    **308.2.2 Refrigerators.** Refrigerators and similar equipment not in operation shall not be discarded, abandoned or stored on *premises* without first removing the doors.

**308.3 Disposal of garbage.** Every *occupant* of a structure shall dispose of garbage in a clean and sanitary manner by placing such garbage in an *approved* garbage disposal facility or *approved* garbage containers.

    **308.3.1 Garbage facilities.** The *owner* of every dwelling shall supply one of the following: an *approved* mechanical food waste grinder in each *dwelling unit;* an *approved* incinerator unit in the structure available to the *occupants* in each *dwelling unit;* or an *approved* leakproof, covered, outside garbage container.

    **308.3.2 Containers.** The *operator* of every establishment producing garbage shall provide, and at all times cause to be utilized, *approved* leakproof containers provided with close-fitting covers for the storage of such materials until removed from the *premises* for disposal.

## SECTION 309
## PEST ELIMINATION

**309.1 Infestation.** All structures shall be kept free from insect and rodent *infestation.* All structures in which insects or rodents are found shall be promptly exterminated by *approved* processes that will not be injurious to human health. After pest elimination, proper precautions shall be taken to prevent reinfestation.

**309.2 Owner.** The *owner* of any structure shall be responsible for pest elimination within the structure prior to renting or leasing the structure.

**309.3 Single occupant.** The *occupant* of a one-family dwelling or of a single-*tenant* nonresidential structure shall be responsible for pest elimination on the *premises.*

**309.4 Multiple occupancy.** The *owner* of a structure containing two or more *dwelling unit*s, a multiple *occupancy,* a *rooming house* or a nonresidential structure shall be responsible for pest elimination in the public or shared areas of the structure and *exterior property*. If *infestation* is caused by failure of an *occupant* to prevent such *infestation* in the area occupied, the *occupant* and *owner* shall be responsible for pest elimination.

**309.5 Occupant.** The *occupant* of any structure shall be responsible for the continued rodent and pest-free condition of the structure.

> **Exception:** Where the *infestations* are caused by defects in the structure, the *owner* shall be responsible for pest elimination.

# CHAPTER 4

# LIGHT, VENTILATION AND OCCUPANCY LIMITATIONS

## SECTION 401
## GENERAL

**401.1 Scope.** The provisions of this chapter shall govern the minimum conditions and standards for light, *ventilation* and space for occupying a structure.

**401.2 Responsibility.** The *owner* of the structure shall provide and maintain light, *ventilation* and space conditions in compliance with these requirements. A person shall not occupy as *owner-occupant*, or permit another person to occupy, any *premises* that do not comply with the requirements of this chapter.

**401.3 Alternative devices.** In lieu of the means for natural light and *ventilation* herein prescribed, artificial light or mechanical *ventilation* complying with the *International Building Code* shall be permitted.

## SECTION 402
## LIGHT

**402.1 Habitable spaces.** Every *habitable space* shall have at least one window of *approved* size facing directly to the outdoors or to a court. The minimum total glazed area for every *habitable space* shall be 8 percent of the floor area of such room. Wherever walls or other portions of a structure face a window of any room and such obstructions are located less than 3 feet (914 mm) from the window and extend to a level above that of the ceiling of the room, such window shall not be deemed to face directly to the outdoors nor to a court and shall not be included as contributing to the required minimum total window area for the room.

> **Exception:** Where natural light for rooms or spaces without exterior glazing areas is provided through an adjoining room, the unobstructed opening to the adjoining room shall be at least 8 percent of the floor area of the interior room or space, but a minimum of 25 square feet (2.33 m²). The exterior glazing area shall be based on the total floor area being served.

**402.2 Common halls and stairways.** Every common hall and stairway in residential occupancies, other than in one- and two-family dwellings, shall be lighted at all times with at least a 60-watt standard incandescent light bulb for each 200 square feet (19 m²) of floor area or equivalent illumination, provided that the spacing between lights shall not be greater than 30 feet (9144 mm). In other than residential occupancies, means of egress, including exterior means of egress, stairways shall be illuminated at all times the building space served by the means of egress is occupied with a minimum of 1 footcandle (11 lux) at floors, landings and treads.

**402.3 Other spaces.** All other spaces shall be provided with natural or artificial light sufficient to permit the maintenance of sanitary conditions, and the safe *occupancy* of the space and utilization of the appliances, equipment and fixtures.

## SECTION 403
## VENTILATION

**403.1 Habitable spaces.** Every *habitable space* shall have at least one openable window. The total openable area of the window in every room shall be equal to at least 45 percent of the minimum glazed area required in Section 402.1.

> **Exception:** Where rooms and spaces without openings to the outdoors are ventilated through an adjoining room, the unobstructed opening to the adjoining room shall be at least 8 percent of the floor area of the interior room or space, but a minimum of 25 square feet (2.33 m²). The *ventilation* openings to the outdoors shall be based on a total floor area being ventilated.

**403.2 Bathrooms and toilet rooms.** Every *bathroom* and *toilet room* shall comply with the *ventilation* requirements for *habitable spaces* as required by Section 403.1, except that a window shall not be required in such spaces equipped with a mechanical *ventilation* system. Air exhausted by a mechanical *ventilation* system from a *bathroom* or *toilet room* shall discharge to the outdoors and shall not be recirculated.

**403.3 Cooking facilities.** Unless *approved* through the certificate of *occupancy*, cooking shall not be permitted in any *rooming unit* or dormitory unit, and a cooking facility or appliance shall not be permitted to be present in the *rooming unit* or dormitory unit.

> **Exceptions:**
>
> 1. Where specifically *approved* in writing by the *code official*.
> 2. Devices such as coffee pots and microwave ovens shall not be considered cooking appliances.

**403.4 Process ventilation.** Where injurious, toxic, irritating or noxious fumes, gases, dusts or mists are generated, a local exhaust *ventilation* system shall be provided to remove the contaminating agent at the source. Air shall be exhausted to the exterior and not be recirculated to any space.

**403.5 Clothes dryer exhaust.** Clothes dryer exhaust systems shall be independent of all other systems and shall be exhausted outside the structure in accordance with the manufacturer's instructions.

> **Exception:** Listed and *labeled* condensing (ductless) clothes dryers.

## SECTION 404
## OCCUPANCY LIMITATIONS

**404.1 Privacy.** *Dwelling units*, hotel units, *housekeeping units*, *rooming units* and dormitory units shall be arranged to provide privacy and be separate from other adjoining spaces.

**404.2 Minimum room widths.** A habitable room, other than a kitchen, shall be a minimum of 7 feet (2134 mm) in any plan dimension. Kitchens shall have a minimum clear passageway of 3 feet (914 mm) between counterfronts and appliances or counterfronts and walls.

**404.3 Minimum ceiling heights.** *Habitable spaces*, hallways, corridors, laundry areas, *bathrooms*, *toilet rooms* and habitable *basement* areas shall have a minimum clear ceiling height of 7 feet (2134 mm).

**Exceptions:**

1. In one- and two-family dwellings, beams or girders spaced a minimum of 4 feet (1219 mm) on center and projecting a maximum of 6 inches (152 mm) below the required ceiling height.

2. *Basement* rooms in one- and two-family dwellings occupied exclusively for laundry, study or recreation purposes, having a minimum ceiling height of 6 feet 8 inches (2033 mm) with a minimum clear height of 6 feet 4 inches (1932 mm) under beams, girders, ducts and similar obstructions.

3. Rooms occupied exclusively for sleeping, study or similar purposes and having a sloped ceiling over all or part of the room, with a minimum clear ceiling height of 7 feet (2134 mm) over a minimum of one-third of the required minimum floor area. In calculating the floor area of such rooms, only those portions of the floor area with a minimum clear ceiling height of 5 feet (1524 mm) shall be included.

**404.4 Bedroom and living room requirements.** Every *bedroom* and living room shall comply with the requirements of Sections 404.4.1 through 404.4.5.

**404.4.1 Room area.** Every living room shall contain at least 120 square feet (11.2 m$^2$) and every bedroom shall contain a minimum of 70 square feet (6.5 m$^2$) and every bedroom occupied by more than one person shall contain a minimum of 50 square feet (4.6 m$^2$) of floor area for each occupant thereof.

**404.4.2 Access from bedrooms.** *Bedrooms* shall not constitute the only means of access to other *bedrooms* or *habitable spaces* and shall not serve as the only means of egress from other *habitable spaces*.

**Exception:** Units that contain fewer than two *bedrooms*.

**404.4.3 Water closet accessibility.** Every *bedroom* shall have access to at least one water closet and one lavatory without passing through another *bedroom*. Every *bedroom* in a *dwelling unit* shall have access to at least one water closet and lavatory located in the same story as the *bedroom* or an adjacent story.

**404.4.4 Prohibited occupancy.** Kitchens and nonhabitable spaces shall not be used for sleeping purposes.

**404.4.5 Other requirements.** *Bedrooms* shall comply with the applicable provisions of this code including, but not limited to, the light, *ventilation*, room area, ceiling height and room width requirements of this chapter; the plumbing facilities and water-heating facilities requirements of Chapter 5; the heating facilities and electrical

receptacle requirements of Chapter 6; and the smoke detector and emergency escape requirements of Chapter 7.

**404.5 Overcrowding.** Dwelling units shall not be occupied by more occupants than permitted by the minimum area requirements of Table 404.5.

**TABLE 404.5**
**MINIMUM AREA REQUIREMENTS**

| SPACE | MINIMUM AREA IN SQUARE FEET | | |
|---|---|---|---|
| | 1-2 occupants | 3-5 occupants | 6 or more occupants |
| Living room[a, b] | 120 | 120 | 150 |
| Dining room[a, b] | No requirement | 80 | 100 |
| Bedrooms | Shall comply with Section 404.4.1 | | |

For SI: 1 square foot = 0.093 m$^2$.

a. See Section 404.5.2 for combined living room/dining room spaces.

b. See Section 404.5.1 for limitations on determining the minimum occupancy area for sleeping purposes.

**404.5.1 Sleeping area.** The minimum occupancy area required by Table 404.5 shall not be included as a sleeping area in determining the minimum occupancy area for sleeping purposes. All sleeping areas shall comply with Section 404.4.

**404.5.2 Combined spaces.** Combined living room and dining room spaces shall comply with the requirements of Table 404.5 if the total area is equal to that required for separate rooms and if the space is located so as to function as a combination living room/dining room.

**404.6 Efficiency unit.** Nothing in this section shall prohibit an efficiency living unit from meeting the following requirements:

1. A unit occupied by not more than one occupant shall have a minimum clear floor area of 120 square feet (11.2 m$^2$). A unit occupied by not more than two *occupants* shall have a minimum clear floor area of 220 square feet (20.4 m$^2$). A unit occupied by three *occupants* shall have a minimum clear floor area of 320 square feet (29.7 m$^2$). These required areas shall be exclusive of the areas required by Items 2 and 3.

2. The unit shall be provided with a kitchen sink, cooking appliance and refrigeration facilities, each having a minimum clear working space of 30 inches (762 mm) in front. Light and *ventilation* conforming to this code shall be provided.

3. The unit shall be provided with a separate *bathroom* containing a water closet, lavatory and bathtub or shower.

4. The maximum number of *occupants* shall be three.

**404.7 Food preparation.** All spaces to be occupied for food preparation purposes shall contain suitable space and equipment to store, prepare and serve foods in a sanitary manner. There shall be adequate facilities and services for the sanitary disposal of food wastes and refuse, including facilities for temporary storage.

# CHAPTER 5

# PLUMBING FACILITIES AND FIXTURE REQUIREMENTS

## SECTION 501
## GENERAL

**501.1 Scope.** The provisions of this chapter shall govern the minimum plumbing systems, facilities and plumbing fixtures to be provided.

**501.2 Responsibility.** The *owner* of the structure shall provide and maintain such plumbing facilities and plumbing fixtures in compliance with these requirements. A person shall not occupy as *owner-occupant* or permit another person to occupy any structure or *premises* which does not comply with the requirements of this chapter.

## SECTION 502
## REQUIRED FACILITIES

**[P] 502.1 Dwelling units.** Every *dwelling unit* shall contain its own bathtub or shower, lavatory, water closet and kitchen sink which shall be maintained in a sanitary, safe working condition. The lavatory shall be placed in the same room as the water closet or located in close proximity to the door leading directly into the room in which such water closet is located. A kitchen sink shall not be used as a substitute for the required lavatory.

**[P] 502.2 Rooming houses.** At least one water closet, lavatory and bathtub or shower shall be supplied for each four *rooming units*.

**[P] 502.3 Hotels.** Where private water closets, lavatories and baths are not provided, one water closet, one lavatory and one bathtub or shower having access from a public hallway shall be provided for each ten *occupants*.

**[P] 502.4 Employees' facilities.** A minimum of one water closet, one lavatory and one drinking facility shall be available to employees.

**[P] 502.4.1 Drinking facilities.** Drinking facilities shall be a drinking fountain, water cooler, bottled water cooler or disposable cups next to a sink or water dispenser. Drinking facilities shall not be located in *toilet rooms* or *bathrooms*.

**[P] 502.5 Public toilet facilities.** Public toilet facilities shall be maintained in a safe sanitary and working condition in accordance with the *International Plumbing Code*. Except for periodic maintenance or cleaning, public access and use shall be provided to the toilet facilities at all times during *occupancy* of the *premises*.

## SECTION 503
## TOILET ROOMS

**[P] 503.1 Privacy.** *Toilet rooms* and *bathrooms* shall provide privacy and shall not constitute the only passageway to a hall or other space, or to the exterior. A door and interior locking device shall be provided for all common or shared *bathrooms* and *toilet rooms* in a multiple dwelling.

**[P] 503.2 Location.** *Toilet rooms* and *bathrooms* serving hotel units, *rooming units* or dormitory units or *housekeeping units*, shall have access by traversing a maximum of one flight of stairs and shall have access from a common hall or passageway.

**[P] 503.3 Location of employee toilet facilities.** Toilet facilities shall have access from within the employees' working area. The required toilet facilities shall be located a maximum of one story above or below the employees' working area and the path of travel to such facilities shall not exceed a distance of 500 feet (152 m). Employee facilities shall either be separate facilities or combined employee and public facilities.

> **Exception:** Facilities that are required for employees in storage structures or kiosks, which are located in adjacent structures under the same ownership, lease or control, shall not exceed a travel distance of 500 feet (152 m) from the employees' regular working area to the facilities.

**[P] 503.4 Floor surface.** In other than *dwelling units*, every *toilet room* floor shall be maintained to be a smooth, hard, nonabsorbent surface to permit such floor to be easily kept in a clean and sanitary condition.

## SECTION 504
## PLUMBING SYSTEMS AND FIXTURES

**[P] 504.1 General.** All plumbing fixtures shall be properly installed and maintained in working order, and shall be kept free from obstructions, leaks and defects and be capable of performing the function for which such plumbing fixtures are designed. All plumbing fixtures shall be maintained in a safe, sanitary and functional condition.

**[P] 504.2 Fixture clearances.** Plumbing fixtures shall have adequate clearances for usage and cleaning.

**[P] 504.3 Plumbing system hazards.** Where it is found that a plumbing system in a structure constitutes a hazard to the *occupants* or the structure by reason of inadequate service, inadequate venting, cross connection, backsiphonage, improper installation, *deterioration* or damage or for similar reasons, the *code official* shall require the defects to be corrected to eliminate the hazard.

## SECTION 505
## WATER SYSTEM

**505.1 General.** Every sink, lavatory, bathtub or shower, drinking fountain, water closet or other plumbing fixture shall be properly connected to either a public water system or to an *approved* private water system. All kitchen sinks, lavatories, laundry facilities, bathtubs and showers shall be supplied with hot or tempered and cold running water in accordance with the *International Plumbing Code*.

**[P] 505.2 Contamination.** The water supply shall be maintained free from contamination, and all water inlets for plumbing fixtures shall be located above the flood-level rim of the fixture. Shampoo basin faucets, janitor sink faucets and other hose bibs or faucets to which hoses are attached and left in place, shall be protected by an approved atmospheric-type vacuum breaker or an approved permanently attached hose connection vacuum breaker.

**505.3 Supply.** The water supply system shall be installed and maintained to provide a supply of water to plumbing fixtures, devices and appurtenances in sufficient volume and at pressures adequate to enable the fixtures to function properly, safely, and free from defects and leaks.

**505.4 Water heating facilities.** Water heating facilities shall be properly installed, maintained and capable of providing an adequate amount of water to be drawn at every required sink, lavatory, bathtub, shower and laundry facility at a minimum temperature of 110°F (43°C). A gas-burning water heater shall not be located in any *bathroom*, *toilet room*, *bedroom* or other occupied room normally kept closed, unless adequate combustion air is provided. An *approved* combination temperature and pressure-relief valve and relief valve discharge pipe shall be properly installed and maintained on water heaters.

### SECTION 506
### SANITARY DRAINAGE SYSTEM

**[P] 506.1 General.** All plumbing fixtures shall be properly connected to either a public sewer system or to an *approved* private sewage disposal system.

**[P] 506.2 Maintenance.** Every plumbing stack, vent, waste and sewer line shall function properly and be kept free from obstructions, leaks and defects.

**[P] 506.3 Grease interceptors.** Grease interceptors and automatic grease removal devices shall be maintained in accordance with this code and the manufacturer's installation instructions. Grease interceptors and automatic grease removal devices shall be regularly serviced and cleaned to prevent the discharge of oil, grease, and other substances harmful or hazardous to the building drainage system, the public sewer, the private sewage disposal system or the sewage treatment plant or processes. All records of maintenance, cleaning and repairs shall be available for inspection by the code official.

### SECTION 507
### STORM DRAINAGE

**[P] 507.1 General.** Drainage of roofs and paved areas, *yards* and courts, and other open areas on the *premises* shall not be discharged in a manner that creates a public nuisance.

# CHAPTER 6

# MECHANICAL AND ELECTRICAL REQUIREMENTS

## SECTION 601
## GENERAL

**601.1 Scope.** The provisions of this chapter shall govern the minimum mechanical and electrical facilities and equipment to be provided.

**601.2 Responsibility.** The *owner* of the structure shall provide and maintain mechanical and electrical facilities and equipment in compliance with these requirements. A person shall not occupy as *owner-occupant* or permit another person to occupy any *premises* which does not comply with the requirements of this chapter.

## SECTION 602
## HEATING FACILITIES

**602.1 Facilities required.** Heating facilities shall be provided in structures as required by this section.

**602.2 Residential occupancies.** Dwellings shall be provided with heating facilities capable of maintaining a room temperature of 68°F (20°C) in all habitable rooms, *bathrooms* and *toilet rooms* based on the winter outdoor design temperature for the locality indicated in Appendix D of the *International Plumbing Code*. Cooking appliances shall not be used, nor shall portable unvented fuel-burning space heaters be used, as a means to provide required heating.

> **Exception:** In areas where the average monthly temperature is above 30°F (-1°C), a minimum temperature of 65°F (18°C) shall be maintained.

**602.3 Heat supply.** Every *owner* and *operator* of any building who rents, leases or lets one or more *dwelling units* or *sleeping units* on terms, either expressed or implied, to furnish heat to the *occupants* thereof shall supply heat during the period from [DATE] to [DATE] to maintain a minimum temperature of 68°F (20°C) in all habitable rooms, *bathrooms* and *toilet rooms*.

**Exceptions:**

1. When the outdoor temperature is below the winter outdoor design temperature for the locality, maintenance of the minimum room temperature shall not be required provided that the heating system is operating at its full design capacity. The winter outdoor design temperature for the locality shall be as indicated in Appendix D of the *International Plumbing Code*.

2. In areas where the average monthly temperature is above 30°F (-1°C) a minimum temperature of 65°F (18°C) shall be maintained.

**602.4 Occupiable work spaces.** Indoor occupiable work spaces shall be supplied with heat during the period from [DATE] to [DATE] to maintain a minimum temperature of 65°F (18°C) during the period the spaces are occupied.

**Exceptions:**

1. Processing, storage and operation areas that require cooling or special temperature conditions.

2. Areas in which persons are primarily engaged in vigorous physical activities.

**602.5 Room temperature measurement.** The required room temperatures shall be measured 3 feet (914 mm) above the floor near the center of the room and 2 feet (610 mm) inward from the center of each exterior wall.

## SECTION 603
## MECHANICAL EQUIPMENT

**603.1 Mechanical appliances.** All mechanical appliances, fireplaces, solid fuel-burning appliances, cooking appliances and water heating appliances shall be properly installed and maintained in a safe working condition, and shall be capable of performing the intended function.

**603.2 Removal of combustion products.** All fuel-burning equipment and appliances shall be connected to an *approved* chimney or vent.

> **Exception:** Fuel-burning equipment and appliances which are *labeled* for unvented operation.

**603.3 Clearances.** All required clearances to combustible materials shall be maintained.

**603.4 Safety controls.** All safety controls for fuel-burning equipment shall be maintained in effective operation.

**603.5 Combustion air.** A supply of air for complete combustion of the fuel and for *ventilation* of the space containing the fuel-burning equipment shall be provided for the fuel-burning equipment.

**603.6 Energy conservation devices.** Devices intended to reduce fuel consumption by attachment to a fuel-burning appliance, to the fuel supply line thereto, or to the vent outlet or vent piping therefrom, shall not be installed unless *labeled* for such purpose and the installation is specifically *approved*.

## SECTION 604
## ELECTRICAL FACILITIES

**604.1 Facilities required.** Every occupied building shall be provided with an electrical system in compliance with the requirements of this section and Section 605.

**604.2 Service.** The size and usage of appliances and equipment shall serve as a basis for determining the need for additional facilities in accordance with NFPA 70. *Dwelling units* shall be served by a three-wire, 120/240 volt, single-phase electrical service having a minimum rating of 60 amperes.

**604.3 Electrical system hazards.** Where it is found that the electrical system in a structure constitutes a hazard to the *occupants* or the structure by reason of inadequate service, improper fusing, insufficient receptacle and lighting outlets, improper wiring or installation, *deterioration* or damage, or for similar reasons, the *code official* shall require the defects to be corrected to eliminate the hazard.

**604.3.1 Abatement of electrical hazards associated with water exposure.** The provisions of this section shall govern the repair and replacement of electrical systems and equipment that have been exposed to water.

**604.3.1.1 Electrical equipment.** Electrical distribution equipment, motor circuits, power equipment, transformers, wire, cable, flexible cords, wiring devices, ground fault circuit interrupters, surge protectors, molded case circuit breakers, low-voltage fuses, luminaires, ballasts, motors and electronic control, signaling and communication equipment that have been exposed to water shall be replaced in accordance with the provisions of the *International Building Code.*

**Exception:** The following equipment shall be allowed to be repaired where an inspection report from the equipment manufacturer or *approved* manufacturer's representative indicates that the equipment has not sustained damage that requires replacement:

1. Enclosed switches, rated a maximum of 600 volts or less;

2. Busway, rated a maximum of 600 volts;

3. Panelboards, rated a maximum of 600 volts;

4. Switchboards, rated a maximum of 600 volts;

5. Fire pump controllers, rated a maximum of 600 voltss;

6. Manual and magnetic motor controllers;

7. Motor control centers;

8. Alternating current high-voltage circuit breakers;

9. Low-voltage power circuit breakers;

10. Protective relays, meters and current transformers;

11. Low- and medium-voltage switchgear;

12. Liquid-filled transformers;

13. Cast-resin transformers;

14. Wire or cable that is suitable for wet locations and whose ends have not been exposed to water;

15. Wire or cable, not containing fillers, that is suitable for wet locations and whose ends have not been exposed to water;

16. Luminaires that are listed as submersible;

17. Motors;

18. Electronic control, signaling and communication equipment.

**604.3.2 Abatement of electrical hazards associated with fire exposure.** The provisions of this section shall govern the repair and replacement of electrical systems and equipment that have been exposed to fire.

**604.3.2.1 Electrical equipment.** Electrical switches, receptacles and fixtures, including furnace, water heating, security system and power distribution circuits, that have been exposed to fire, shall be replaced in accordance with the provisions of the *International Building Code.*

**Exception:** Electrical switches, receptacles and fixtures that shall be allowed to be repaired where an inspection report from the equipment manufacturer or *approved* manufacturer's representative indicates that the equipment has not sustained damage that requires replacement.

## SECTION 605
## ELECTRICAL EQUIPMENT

**605.1 Installation.** All electrical equipment, wiring and appliances shall be properly installed and maintained in a safe and *approved* manner.

**605.2 Receptacles.** Every *habitable space* in a dwelling shall contain at least two separate and remote receptacle outlets. Every laundry area shall contain at least one grounded-type receptacle or a receptacle with a ground fault circuit interrupter. Every *bathroom* shall contain at least one receptacle. Any new *bathroom* receptacle outlet shall have ground fault circuit interrupter protection. All receptacle outlets shall have the appropriate faceplate cover for the location.

**605.3 Luminaires.** Every public hall, interior stairway, *toilet room*, kitchen, *bathroom*, laundry room, boiler room and furnace room shall contain at least one electric luminaire. Pool and spa luminaries over 15 V shall have ground fault circuit interrupter protection.

**605.4 Wiring.** Flexible cords shall not be used for permanent wiring, or for running through doors, windows, or cabinets, or concealed within walls, floors, or ceilings.

## SECTION 606
## ELEVATORS, ESCALATORS AND DUMBWAITERS

**606.1 General.** Elevators, dumbwaiters and escalators shall be maintained in compliance with ASME A17.1. The most current certificate of inspection shall be on display at all times within the elevator or attached to the escalator or dumbwaiter, be available for public inspection in the office of the building *operator* or be posted in a publicly conspicuous location *approved* by the *code official*. The inspection and tests shall be performed at not less than the periodic intervals listed in ASME A17.1, Appendix N, except where otherwise specified by the authority having jurisdiction.

**606.2 Elevators.** In buildings equipped with passenger elevators, at least one elevator shall be maintained in operation at all times when the building is occupied.

> **Exception:** Buildings equipped with only one elevator shall be permitted to have the elevator temporarily out of service for testing or servicing.

## SECTION 607
## DUCT SYSTEMS

**607.1 General.** Duct systems shall be maintained free of obstructions and shall be capable of performing the required function.

# CHAPTER 7

# FIRE SAFETY REQUIREMENTS

## SECTION 701
## GENERAL

**701.1 Scope.** The provisions of this chapter shall govern the minimum conditions and standards for fire safety relating to structures and exterior *premises*, including fire safety facilities and equipment to be provided.

**701.2 Responsibility.** The *owner* of the *premises* shall provide and maintain such fire safety facilities and equipment in compliance with these requirements. A person shall not occupy as *owner-occupant* or permit another person to occupy any *premises* that do not comply with the requirements of this chapter.

## SECTION 702
## MEANS OF EGRESS

**[F] 702.1 General.** A safe, continuous and unobstructed path of travel shall be provided from any point in a building or structure to the *public way*. Means of egress shall comply with the *International Fire Code*.

**[F] 702.2 Aisles.** The required width of aisles in accordance with the *International Fire Code* shall be unobstructed.

**[F] 702.3 Locked doors.** All means of egress doors shall be readily openable from the side from which egress is to be made without the need for keys, special knowledge or effort, except where the door hardware conforms to that permitted by the *International Building Code*.

**[F] 702.4 Emergency escape openings.** Required emergency escape openings shall be maintained in accordance with the code in effect at the time of construction, and the following. Required emergency escape and rescue openings shall be operational from the inside of the room without the use of keys or tools. Bars, grilles, grates or similar devices are permitted to be placed over emergency escape and rescue openings provided the minimum net clear opening size complies with the code that was in effect at the time of construction and such devices shall be releasable or removable from the inside without the use of a key, tool or force greater than that which is required for normal operation of the escape and rescue opening.

## SECTION 703
## FIRE-RESISTANCE RATINGS

**[F] 703.1 Fire-resistance-rated assemblies.** The required fire-resistance rating of fire-resistance-rated walls, fire stops, shaft enclosures, partitions and floors shall be maintained.

**[F] 703.2 Opening protectives.** Required opening protectives shall be maintained in an operative condition. All fire and smokestop doors shall be maintained in operable condition. Fire doors and smoke barrier doors shall not be blocked or obstructed or otherwise made inoperable.

## SECTION 704
## FIRE PROTECTION SYSTEMS

**[F] 704.1 General.** All systems, devices and equipment to detect a fire, actuate an alarm, or suppress or control a fire or any combination thereof shall be maintained in an operable condition at all times in accordance with the *International Fire Code*.

**[F] 704.1.1 Automatic sprinkler systems.** Inspection, testing and maintenance of automatic sprinkler systems shall be in accordance with NFPA 25.

**[F] 704.2 Smoke alarms.** Single- or multiple-station smoke alarms shall be installed and maintained in Group R or I-1 occupancies, regardless of *occupant* load at all of the following locations:

1. On the ceiling or wall outside of each separate sleeping area in the immediate vicinity of *bedrooms*.

2. In each room used for sleeping purposes.

3. In each story within a *dwelling unit*, including *basements* and cellars but not including crawl spaces and uninhabitable attics. In dwellings or *dwelling units* with split levels and without an intervening door between the adjacent levels, a smoke alarm installed on the upper level shall suffice for the adjacent lower level provided that the lower level is less than one full story below the upper level.

**[F] 704.3 Power source.** In Group R or I-1 occupancies, single-station smoke alarms shall receive their primary power from the building wiring provided that such wiring is served from a commercial source and shall be equipped with a battery backup. Smoke alarms shall emit a signal when the batteries are low. Wiring shall be permanent and without a disconnecting switch other than as required for overcurrent protection.

**Exception:** Smoke alarms are permitted to be solely battery operated in buildings where no construction is taking place, buildings that are not served from a commercial power source and in existing areas of buildings undergoing alterations or repairs that do not result in the removal of interior wall or ceiling finishes exposing the structure, unless there is an attic, crawl space or basement available which could provide access for building wiring without the removal of interior finishes.

**[F] 704.4 Interconnection.** Where more than one smoke alarm is required to be installed within an individual *dwelling unit* in Group R or I-1 occupancies, the smoke alarms shall be interconnected in such a manner that the activation of one alarm will activate all of the alarms in the individual unit. Physical interconnection of smoke alarms shall not be required where listed wireless alarms are installed and all alarms sound upon activation of one alarm. The alarm shall be clearly audible in all *bedrooms* over background noise levels with all intervening doors closed.

**Exceptions:**

1. Interconnection is not required in buildings which are not undergoing alterations, repairs or construction of any kind.

2. Smoke alarms in existing areas are not required to be interconnected where alterations or repairs do not result in the removal of interior wall or ceiling finishes exposing the structure, unless there is an attic, crawl space or *basement* available which could provide access for interconnection without the removal of interior finishes.

# CHAPTER 8

# REFERENCED STANDARDS

This chapter lists the standards that are referenced in various sections of this document. The standards are listed herein by the promulgating agency of the standard, the standard identification, the effective date and title and the section or sections of this document that reference the standard. The application of the referenced standards shall be as specified in Section 102.7.

## ASME
American Society of Mechanical Engineers
Three Park Avenue
New York, NY 10016-5990

| Standard reference number | Title | Referenced in code section number |
|---|---|---|
| A17.1/CSA B44—2007 | Safety Code for Elevators and Escalators | 606.1 |

## ASTM
ASTM International
100 Barr Harbor Drive
West Conshohocken, PA 19428-2959

| Standard reference number | Title | Referenced in code section number |
|---|---|---|
| F 1346—91 (2003) | Performance Specifications for Safety Covers and Labeling Requirements for All Covers for Swimming Pools, Spas and Hot Tubs | 303.2 |

## ICC
International Code Council
500 New Jersey Avenue, NW
6th Floor
Washington, DC 20001

| Standard reference number | Title | Referenced in code section number |
|---|---|---|
| IBC—12 | International Building Code® | 102.3, 201.3, 401.3, 702.3 |
| IEBC—12 | International Existing Building Code® | 305.1.1, 306.1.1 |
| IFC—12 | International Fire Code® | 201.3, 604.3.1.1, 604.3.2.1, 702.1, 702.2, 704.1, 704.2 |
| IFGC—12 | International Fuel Gas Code® | 102.3 |
| IMC—12 | International Mechanical Code® | 102.3, 201.3 |
| IPC—12 | International Plumbing Code® | 201.3, 505.1, 602.2, 602.3 |
| IRC—12 | International Residential Code® | 201.3 |
| IZC—12 | International Zoning Code® | 102.3, 201.3 |

## NFPA
National Fire Protection Association
1 Batterymarch Park
Quincy, MA 02269

| Standard reference number | Title | Referenced in code section number |
|---|---|---|
| 25—11 | Inspection, Testing and Maintenance of Water-Based Fire Protection Systems | 704.1.1 |
| 70—11 | National Electrical Code | 102.4, 201.3, 604.2 |

# APPENDIX A

# BOARDING STANDARD

*The provisions contained in this appendix are not mandatory unless specifically referenced in the adopting ordinance.*

## A101
## GENERAL

**A101.1 General.** All windows and doors shall be boarded in an *approved* manner to prevent entry by unauthorized persons and shall be painted to correspond to the color of the existing structure.

## A102
## MATERIALS

**A102.1 Boarding sheet material.** Boarding sheet material shall be minimum $^1/_2$-inch (12.7 mm) thick wood structural panels complying with the *International Building Code*.

**A102.2 Boarding framing material.** Boarding framing material shall be minimum nominal 2-inch by 4-inch (51 mm by 102 mm) solid sawn lumber complying with the *International Building Code*.

**A102.3 Boarding fasteners.** Boarding fasteners shall be minimum $^3/_8$-inch (9.5 mm) diameter carriage bolts of such a length as required to penetrate the assembly and as required to adequately attach the washers and nuts. Washers and nuts shall comply with the *International Building Code*.

## A103
## INSTALLATION

**A103.1 Boarding installation.** The boarding installation shall be in accordance with Figures A103.1(1) and A103.1(2) and Sections A103.2 through A103.5.

**A103.2 Boarding sheet material.** The boarding sheet material shall be cut to fit the door or window opening neatly or shall be cut to provide an equal overlap at the perimeter of the door or window.

**A103.3 Windows.** The window shall be opened to allow the carriage bolt to pass through or the window sash shall be removed and stored. The 2-inch by 4-inch (51 mm by 102 mm) strong back framing material shall be cut minimum 2 inches (51 mm) wider than the window opening and shall be placed on the inside of the window opening 6 inches minimum above the bottom and below the top of the window opening. The framing and boarding shall be predrilled. The assembly shall be aligned and the bolts, washers and nuts shall be installed and secured.

**A103.4 Door walls.** The door opening shall be framed with minimum 2-inch by 4-inch (51 mm by 102 mm) framing material secured at the entire perimeter and vertical members at a maximum of 24 inches (610 mm) on center. Blocking shall also be secured at a maximum of 48 inches (1219 mm) on center vertically. Boarding sheet material shall be secured with screws and nails alternating every 6 inches (152 mm) on center.

**A103.5 Doors.** Doors shall be secured by the same method as for windows or door openings. One door to the structure shall be available for authorized entry and shall be secured and locked in an *approved* manner.

## A104
## REFERENCED STANDARDS

IBC—12   International Building Code   A102.1, A102.2, A102.3

3/8 inch carriage bolts.
Bolts shall be long enough to extend from the
exterior plywood through the interior plywood
and strong backs and fastened from the interior
with a nut.

12in

6 in

1/2 inch CDX Plywood or
Performance rated OSB

2" x 4" Strong Backs

Window Frame

2" x 4" Strong Backs

3/8 inch carriage bolts.
Bolts shall be long enough to extend from the
exterior plywood through the interior plywood
and strong backs and fastened from the interior
with a nut.

**FIGURE A103.1(1)**
**BOARDING OF DOOR OR WINDOW**

1/2 inch CDX plywood or performance rated OBS shall be secured
to header, base plate, studs, stiles, and edge blocking using alternate
screws and nails at a maximum of 6 inch OC

2" x 4" header

2" x 4" stile

Door wall frame

2" x 4" base
plate

2" x 4" edge blocking
either horizontally or
vertically along edge
of each sheet of plywood
or OSB

2" x 4" studs
Spaced 24" on center

**FIGURE A103.1(2)**
**BOARDING OF DOOR WALL**

# INDEX

## A

**ACCESS**

**ADJACENT**

**ADMINISTRATION**

**AGENT (See also OPERATOR, OWNER)**

**AIR**

**AISLES**

**ALTERATION**

**ANCHOR**

**APPEAL**

**APPLIANCE**

**APPLICABILITY**

**APPROVAL**

**APPROVED**

**ARTIFICIAL**

**AUTOMOBILE**

**AWNING**

## B

**BALCONY**

**BASEMENT**

**BATHROOM**

**BATHTUB**

**EDITORIAL CHANGES – SECOND PRINTING**

Page 25, Section [F] 704.2: now reads . . .**[F] 704.2 Smoke alarms.** Single- or multiple-station smoke alarms shall be installed and maintained in Group R or I-1 occupancies, regardless of occupant load at all of the following locations:

1. On the ceiling or wall outside of each separate sleeping area in the immediate vicinity of *bedrooms*.

2. In each room used for sleeping purposes.

3. In each story within a dwelling unit, including basements and cellars but not including crawl spaces and uninhabitable attics. In dwellings or dwelling units with split levels and without an intervening door between the adjacent levels, a smoke alarm installed on the upper level shall suffice for the adjacent lower level provided that the lower level is less than one full story below the upper level.

Page 25, Section [F] 704.3: lines 1 and 2 now reads . . .**[F] 704.3 Power source.** In Group R or I-1 occupancies, single-station smoke alarms shall receive their primary power

Page 25, Section [F] 704.4: now reads . . .**[F] 704.4 Interconnection.** Where more than one smoke alarm is required to be installed within an individual *dwelling unit* in Group R or I-1 occupancies, the smoke alarms shall be interconnected in such a manner that the activation of one alarm will activate all of the alarms in the individual unit. Physical interconnection of smoke alarms shall not be required where listed wireless alarms are installed and all alarms sound upon activation of one alarm. The alarm shall be clearly audible in all bedrooms over background noise levels with all intervening doors closed.

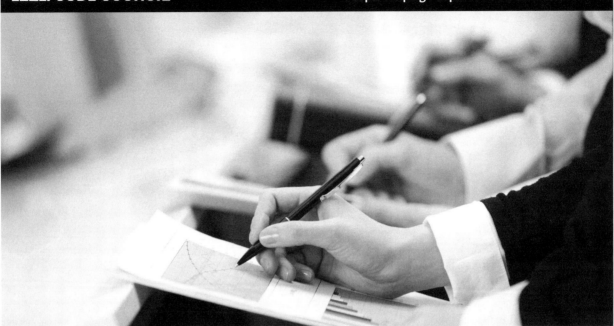

# When it comes to code education, ICC has you covered.

ICC publishes building safety, fire prevention and energy efficiency codes that are used in the construction of residential and commercial buildings. Most U.S. cities, counties, and states choose the I-Codes based on their outstanding quality.

ICC also offers the highest quality training resources and tools to properly apply the codes.

## TRAINING RESOURCES

- **Customized Training:** Training programs tailored to your specific needs.
- **Institutes:** Explore current and emerging issues with like-minded professionals.
- **ICC Campus Online:** Online courses designed to provide convenience in the learning process.
- **Webinars:** Training delivered online by code experts.
- **ICC Training Courses:** On-site courses taught by experts in their field at select locations and times.

## TRAINING TOOLS

- **Online Certification Renewal Update Courses:** Need to maintain your ICC certification? We've got you covered.
- **Training Materials:** ICC has the highest quality publications, videos and other materials.

**For more information on ICC training, visit http://www.iccsafe.org/Education or call 1-888-422-7233, ext. 33818.**